JOHN CONNELL's work has been published in *Granta*'s *New Irish Writing* issue. He lives on his family farm, Birchview, in County Longford.

'An unexpected triumph: a brutally realistic account of mud and slime and relentless rain and incredibly hard labour . . . A satisfying, powerful mix. One of the joys of this book is the prose, its clean plainness offset by the glorious cadences of Irish speech, which are also those of the Bible and prayer book' *Daily Mail*

'A brooding, powerful memoir . . . The strength and originality of this book clearly lies in Connell's searingly honest account of rebuilding his life' *Guardian*

'*The Cow Book* starts well . . . and gets better. This is an important slice of living, breathing agricultural history written with absolute honesty by an insider who has the rare quality of being able to see it from the outside too' Rosamund Young, author of *The Secret Life of Cows*

'A gorgeous read, full of warmth, truth and tentative wonder. John Connell has written "an elegy to the nature I know", but this book is an elegy to so much more – to art and myth and sorrow and longing' Sara Baume, author of *Spill Simmer Falter Wither* and *A Line Made By Walking*

'John Connell writes about the connection between people and land in a way that goes beyond mere affection. For him, farming is hard graft and yet a spiritual process too that binds him to family, nationhood, language and myth' Andrew Michael Hurley, author of *The Loney*

'A beautiful book. The author's voice is as honest and tender as his eye is clear and sharp. The apparent simplicity of its subject matter belies the poetic power of the writing. This poignant book is about family, about our ties to the land, and about the cycles of life and death that mark our days' Zia Haider Rahman, author of *In the Light of What We Know*

The Cow Book

A STORY OF LIFE ON AN IRISH FAMILY FARM

JOHN CONNELL

GRANTA

0585
0599

593

Granta Publications, 12 Addison Avenue, London, W11 4QR
First published in Great Britain by Granta Books, 2018
This edition published by Granta Books, 2019

A CIP catalogue record for this book is available from the British Library.

9 8 7 6 5 4

ISBN 978 1 78378 418 9 (paperback)
ISBN 978 1 78378 419 6 (ebook)

Typeset by Avon DataSet Ltd, Arden Court, Alcester, Warwickshire
Printed and bound by CPI Group (UK) Ltd, Croydon, CR0 4YY

www.granta.com

For Mary, Mick & John
With thanks to Simon
And David Malouf, my dear friend and teacher

The tracks of cattle to a drinking-place,
A green stone lying sideways in a ditch,
Or any common sight, the transfigured face
Of a beauty that the world did not touch.

Patrick Kavanagh, 'A Christmas Childhood'

'Live in each season as it passes; breathe the air, drink the drink, taste the fruit, and resign yourself to the influence of each.'

Henry David Thoreau, *Journal*, 23 August 1853

JANUARY

Beginnings

I'm twenty-nine and I've never delivered a calf myself. But that's all about to change because I've got my arms in her passage and I'm trying to find the new calf's feet.

As a farmer's son, I've birthed calves aplenty, but always as the helper, holding a cow's tail up or pulling the calf out at the last moment. My father has been in charge of the calving for twenty-five years, and when he wasn't my brother took over, but now it's me.

I'm home again in rural Ireland, back from being an emigrant, here to write a novel, to try and make it as a writer, and, in exchange for a roof over my head, it's been agreed that I will help out on the farm. There's a lot tied up in this birth for me, much more than the cow knows.

The red cow moves suddenly and lets me know her strength and power. I must be quick. I must get the ropes around the calf's feet, slide them above the hock and pull them tight. The amniotic fluids wet my hands, my arms, and I remember now the talk I have heard other men say, that your hand gets weak after a time, that the clasp of her vaginal embrace takes the power from it. I must be quick lest the calf die.

I think for a moment that I am glad I am alone and doing

this myself. I could call for help. I could, but then I would still not be tested, would still not be able to prove that I can do this myself. I will call no one. I grasp the first foot and slip the rope over his hock.

I had been watching the red cow all night and could sense that she was going to calve. She was sick, as Mam calls it, patrolling her pen, not eating her silage or nuts or water, and then I saw that her passage was broken down, sausage-like.

Da – my father, Tom – is away. He is at the sheep mart with his brother Davy. As older men, they have found each other again as friends. They go to the sheep sales together every week now. Sheep are a new thing for us – we have only been keeping them three years. We have a flock now, as does Davy. He and Da buy and sell the animals for fun; it gives them a hobby and something to do together, and above all it makes my father happy, for he is a sociable man. I never begrudge him these trips, for I know they do him good and give him a break from the farm. I know too that he won't be in the mood to fight after them. That is the most important thing, for we are trying, this calving season, not to fight, and so far it is working. I cannot say that we are friends yet, but a respect has come between us that was never there before. It is a small and delicate thing, still fragile.

I have the other foot. I take the second rope from the side of the gate and slide it onto the calf's leg. It slips and falls and I curse, and now I think perhaps I do need help, but it is too late. To wait might mean death and then I would be called a fool for trying on my own and there would be a huge row. No, I must focus. The cow has nearly finished the nuts I gave her to keep her calm. When they are gone, she will remember her distress and begin to thrash and kick again, and then the job will be all the harder.

I stoop low, take the rope and turn to my work again. The rope is now on the second foot. I pull gently, but the calf is big: he will not come like this; I will need the jack. I take the mechanical wrench, placing it on the cow's hips, and hook the ropes into the slots and begin to winch.

I must do this right, I tell myself, but I have seen it done so many times I know my actions. I must jack, then lever down to bring the calf out. The biggest pull will be his head, and once I have that out the rest should follow, except the hips, which can sometimes be trouble. I winch the jack five times and hear the sprocket chime out in the quiet shed. I pull down and as I do the cow bellows low in a noise I don't recognize, a noise of pain and strangeness.

'There, there,' I say, clucking to her. I let off the pressure and jack once more and feel the sprockets turn and the ratchet move up the teeth. The legs emerge more fully now, but still no head, and so I lever down again and I can see his nose; it looks so flat, perhaps his head is squashed. The cow bellows low again and I feel her feet tremble.

'Don't go down on me,' I say, and let the pressure off once more and she stands to again and we repeat our chores. Her contractions push the calf as much as she can, but he is beyond contractions now, for he is too big and our job is at a point where it cannot be undone.

I jack once more and the cow roars. I am sure Mam will waken now, for she is such a light sleeper. It was she I turned to for advice tonight to make sure I did not take the calf too soon. She has known cows all her life and is wiser than us all with them, but again I remind myself: this is my job now.

I pray, or at least I think I do. The head emerges and I have no time to thank God, for I must jack with all my might and keep going, for the cow could give way and if she goes down, the calf might die. I see his tongue wag and I know he is alive and I pull still stronger now, though my arm is growing tired. I jack and jack and he is emerging now, fluid and strong, and he is red like his mother, with a white sketch on his face. He is the son of our stock bull, of that I am sure, for I can see the old bull's face in his.

The calf's hips are big and the winch of the jack is at its end. The cow is in distress and I remember now Da's words to twist the calf to bring the hips out, so I do and I take him now in both my arms and the adrenaline is such I do not feel his weight. I carry him to the fresh bedding, jack, ropes and all. I must move quick still, for we have lost calves with fluid on their lungs before.

I pour water in his ears and he shakes his head and comes to moving life, and I smile with relief. But then he coughs and I can hear the fluid, so I take a breathing tube with a mask on its end and fit it over the calf's muzzle. You extend the pump and its vacuum pulls the fluid up and, in theory, the calf should cough up the fluid. I do this three times, but the fluid does not come up and he begins to wheeze. I cannot lose him. I pick him up with a roar and carry him over to the gate and sling him across it.

I have seen this done before, but the calf has always been

lifted by two men, so I know that I must have found new strength. I massage his lungs, and give him a slap and soon I see the mucus emerge. He lifts his head and I know that he is won. I release him down into my arms and carry him back to the fresh bedding, alive and safe.

I disinfect his navel, take a small breather and walk to the farm kitchen. There is blood on my arms and face, but it is a pleasing blood, the blood of life. I rinse my hands in a barrel of water. The frost has come and the water is cold and it stings.

My job is not done, though, for I need to get the calf to feed. When a cow gives birth her milk is of a special kind that we call 'beestings'. This colostrum is thick and yellow and the calves must have it straight away, for it keeps them alive and gives them the necessary antibodies to ward off infections and sickness. The first few hours in a calf's life are its most important; if these things are not done – if he is not fed, his navel not treated – any number of things could kill him. Pneumonia is a plague to us farmers, it has killed so many calves; scour too has taken its toll of death.

We always tube-feed our newborn calves. It means we know they get milk into them, and you have ruled out a whole lot of checking on them later in the night. The stomach tube is a plastic pipe connected to a bag. The pipe is inserted down the calf's throat and the milk goes directly to his stomach. This is a dangerous job, for if the pipe is inserted incorrectly, it could go into the lung and kill him as soon as the milk starts flowing. First, I must milk the cow. I wet my hands and take her tit between my fingers. I strug it and, after two or three pulls, the milk squirts out. I massage it now, as I have seen my father do, and find the natural rhythm and soon my jug is filling. The beestings is thick and warm like custard. The cow makes to kick but I am too quick and rescue my jug.

'Easy, girl,' I say, and talk sweet nonsense to her. Her hard work is done and I shall not abuse her now.

The calf rustles in the straw and she turns and shouts to him and he cries a small, plaintive cry, as much to say, I am alive.

I steady myself again and move on to her next tit and strug and milk and soon the milk is squirting into my jug, singing in the age-old sound of milk pouring onto itself. It is a sound all farmers know. It is the sound of my childhood and my parents' and theirs before them. I think now that this family that I am part of has been doing this for so long – so many sleepless late nights spent milking in sheds and barns and, who knows, before my grandfather's time, perhaps on the cottage floor by the fire in the old house. The times have changed, but not the animals, and not our actions.

My jug is now overflowing, so I pour it into the stomach tube bag. I take another jug and by the time it is full I have broken the seal of all four tits and they will be easier for the calf to suck. This cow's teats are not so big, which is a good thing, for I have seen cows with big, long, dangly ones that calves struggle to hold in their mouths, and that creates its own problems.

The stomach tube bag is full now, so I must begin to feed him. This too is a first for me and I must be careful. I sit lightly atop the calf's back so he cannot struggle. Then I lift his head, prise open his mouth with one hand and insert the tube with the other.

'That's it,' I say, and I do not know if I am talking to the calf or to myself, but together we are doing this and slowly now the pipe moves down his throat. I think I have avoided his lungs, so I lift the bag into the air and watch the colostrum drain away down the tube and inside him, and I know the end of the job is near.

'Thanks be to God,' I say, in the way that I have heard others speak, and I am glad of this automatic phrase, for it is the truest saying that I can find in this moment.

I release the calf and sit back now, resting against the whitewashed wall of the calving house. And with that the half door opens and it is my father, bright and smiling.

'She calved.'

'Just pulled him there.'

'You should have called your brother.'

'Can't always be calling people,' I say. 'Sometimes you just have to do it yourself.'

'Sometimes you do,' he says, and smiles.

And I know now that something has happened. I've passed a test of some kind and I am glad. He opens the half door and walks in. He is in his jobbing coat, which is his blue velvety coat for the mart. My uncle Davy follows behind, along with my young cousin, Jack.

'There's money being made here,' says Davy, and we laugh.

I stand up now and they admire the calf. He is a fine wee bull.

The boys leave to go to their house and it is just Da and me again.

'Had ya a good night?' I ask.

'Grand,' he says.

'Any sheep?'

'No, but it was good to look.'

I unloose the cow and leave her and her newborn to each other. She licks him, gentle and soft, despite her size. Nature will do the rest. I am twenty-nine, but I feel so much older this night.

Summers Past

Manhood is an important thing in this land. Farming gives us our sense of it, our understanding of ourselves. My father is a manly man – it is something I have always admired in him.

As a *gasoon* – a boy – I remember summers past in the long ago when he and his brothers Mick and John worked to save the hay. We were in the upper ground and the day was fine. They were the age I am now, and I marvelled as they pulled and threw the square bales of hay high atop the trailer, working in unison like men on the line, like soldiers on a march.

They joked and sang and chatted and the work passed quickly. I envied them their strength then.

They worked as poets of the field, bards of the land; their

common speech had a musicality I have sought to emulate in my adult years.

Turning now, coming up twenty years later, it is me who throws the hay and they who watch in their different forms.

My father and his brothers: the first great farmers I ever knew.

The Farm

Our farmland is flat and thick with hedges and trees. The ground is average, though we have made it good with our work and sweat. It was all bog and rush when my parents came to live here. Da built the house as a young man and slowly, as life moved forward, together he and my mother built the farm. We have our acres around the house, as well as our lands of Esker, Ruske's and Clonfin, which are across and down the roads. A stream flows by the house and by the end of the fields it joins with the Camlin, the biggest river in the county of Longford. I still remember when we built the hay shed and the holes for the iron girders filled with a foot of water and I found frogs swimming in them. Many neighbours came to help us erect the shed, like an American barn raising. Here we call it a *meitheal* – a gathering of men to work together, like in the old times. We have been farming in this place nearly thirty years now, starting off with just three cows, which have grown to a large herd. I remember those three cows still with fondness, as a city man might remember a departed pet dog.

We are in the depths of winter. It's been the wettest January on record and storms have battered and blown our sheds and fields and rivers. In the west of the country people are flooded, and I've watched them cry on the news. I met an agriculture salesman in the feed store who had come from Galway, and he told me they had rowed boats over fields, over the height of gates. I cannot imagine our fields being so drowned.

Our fields have names, but I do not think this strange, for all our neighbours have names for their fields too. They have been passed on from generation to generation. When Da and Mam bought Ruske's ten years ago, after old Robin died, the first thing they did was to find out their names. There was the Crab Apple field, the Potato field, and the Meadow. They were English names, for the Ruske's land had been owned by the Hamiltons once, and they had been settlers in the time of the plantations of Queen Elizabeth I. We no longer know what the Gaelic names for the fields were; they are gone. Once on the radio I heard a well-known sports commentator from Kerry speak of the childhood fields of his home: the names were in Irish and sounded beautiful.

The fields here are old and have known my people, the Connells, for a long time, first as tenants to an English lord, then as owners. We have walked and worked their knolls and nooks. There have been other families, other bloodlines, in this place in the townland of Soran, but many have moved away or died out. Farming is a walk with survival, with death over our shoulder, sickness to our left, the spirit to our right and the joy of new life in front. It is a cross of creation, like the sign of God we were taught in school.

There are three sheds in the yard, each built at different times as the farm grew. Once our old accountant tried to get my parents to invest in apartments in a tourist town but they refused. The land is what we know, they said. It sustains us, enriches us, the land is our living and we know no other way. Birchview is the name of this place. It is my home.

Mornings

Mother is the first to rise. I do not know anyone who works quite as hard as her. She is our gatekeeper, for between the hours of 3 and 6 a.m. no one is awake, unless there is a birth, and so her morning watch at six is the first of the new day and will tell us what to expect.

The farm is not her job; she has a Montessori school and day-care centre at the back of the house, but she loves the farm all the same.

I must admit that there have been mornings when I have cursed her waking of me, but then I am reminded that the cows or sheep have no one else to help them in their hours of need or distress. Mother is their voice.

I rise at the same time each morning, have a cup of coffee and a bowl of porridge and make my way to the farmyard. En route I check Facebook to see if there are messages from my girlfriend, Vivian. We're operating on different schedules, for she is in Australia and so her night is my day. We catch each other in the mornings like ships passing, tell each other our news, how I slept, how her day was.

It is still dark as I walk across the yard. The frost has not come, but it is cold and I blow upon my hands to warm them. I do not wear gloves, for we farmers do not wear gloves – I think it might be seen as weakness.

My first job of the day is to let the dog out.

His name is Vinny, he's a pup, and I've been training him these last few weeks. I've never trained a dog before, so cannot gauge either my work or his discipline, and yet he does as he is told now, which is just as well for Vinny, because he was nearly a goner.

I blame myself, for I started it. The sheep had been lambing for a few weeks when Vinny arrived on the farm. He was bright and young and great company for us as we checked on the animals in the night. It must have been the third or fourth lambing when I threw him the afterbirth, which is the placenta of the mother. He set to it and chewed as if it were the finest tartare. I wondered briefly if it was the right thing to do, but he did not bother the lambs or sheep so it seemed that it was OK, and so the practice began. I would deliver a lamb and throw Vinny the afterbirth and, being the good dog, he was would eat it promptly.

It was six in the morning when Mam woke me.

'Vinny's eating one of the new lambs,' she cried.

I did not see the attack but I saw the lamb afterwards, his ears chewed red and bloody, his legs clawed. He was shaken but he was alive. Mother had wrestled the two apart and hushed the dog away.

I treated the lamb's cuts with iodine and put him back with his mother.

We worked out as best we could what had happened. The lamb had been born in the night and had escaped its pen, still covered in birth fluids. Vinny had thought it another placenta – lunch, a mobile lunch – but I do not think he minded that and he would have chewed the lamb up happily, not out of violence towards the creature but in getting his treat of meat.

'He's got no sense,' Mam said.

'Do you think he meant it?' I asked.

'I don't think so.'

I scolded Vinny then, for the sight of the lamb's ears angered me so.

When my father came out to the yard he too scolded the dog and then, carrying him by the scruff of his neck, placed him in his dog house. He stayed there for a time, whilst we decided what to do with him.

My father is not a dog person, so it was odd that he was the one who had bought Vinny and even given him his name. After a few days of thinking, he took the dog in the jeep to give him a new home with a neighbour by the hill farm of Clonfin, some twenty minutes away. I did not say anything then. I now think that cowardice comes in different forms; sometimes it is as simple as silence.

In the end we decided against getting rid of Vinny. Mam and I called Da and asked that he bring the dog home. It came down to faith, we reasoned – faith that the dog could be trained to unlearn bad habits. There is nothing more dangerous to a farmer than a bad dog, for they can kill sheep, drive cattle to madness and break your heart.

Like Pontius Pilate, my father absolved himself of all responsibility for the dog that day. He put Vinny into my care, and there he has stayed. I have not fed him an afterbirth since; I do not wish to tempt the dog and I do not wish to beat him. Vinny is a reformed character now, remembering his brief punishment all too well. There is a learning in everything on a farm, something that I am growing to know as I age.

Vinny keeps to my heels now as I walk over to the cowsheds. The cows roar at the sight of me, demanding their silage, which is a fermented grass that was harvested in the summer. I was not here for the harvest. Some mornings it smells sweetly and reminds us of summer and meadows. This morning it has no smell. My father and I mostly feed the cows together, but some mornings I do it alone. Today he is having a sleep in. This suits me fine, I think, so he cannot nitpick my way of doing things, and, after his decades of early mornings, I like to give him the odd morning off.

The cows eat a bale of silage each morning and another in the evening. In the past we had to cut open the bales and use our long-handled pitch forks – or grapes, as we call them – to manually distribute the feed amongst the cows, but this winter my brother gifted my father an automatic feeder, which rolls and chews up a round bale and spits it out in a neat row, neater than I could ever do by hand. It saves waste and time. The cows low until they are fed and then set to eating and chewing. It takes two hours each morning to feed all the stock, and only when every animal has food in front of it does the bellowing stop.

In the lower shed the weanlings are being fattened. They will go for slaughter in a few weeks, so we keep food in front of them at all times. There is a young red bull that I have admired. He is strong and muscular and will bring a good price. He nearly killed me one day while I was cleaning the shed, but I have forgiven him that, and I suppose he has forgiven me for the beating I gave him in return. I know it was only curiosity that drove him to try and headbutt me. Cattle have their own personalities. Like dogs, or people for that matter, some cows are nice, some are bad, some are sly and some are just lazy. Their temperaments vary and their moods change. I have seen the kindest cow bully one of its fellows and the angriest bull play with young calves. There is no racism in cows, and the different breeds and colours all get along together.

After everything is fed, I check the hens for eggs and give them feed and water. They are laying well at the moment and I praise them, as I do all the animals when they are good. Some mornings the eggs are still warm; I imagine the poached or boiled egg I will eat later and smile, for it will be a nice treat. But before that: mucking out.

The cows that have recently calved are in the single pens and they must be cleaned and bedded. Cows don't have the sense to void in a corner, or to bury or cover over their waste, and so the houses grow filthy. I don't like to see them stand in shite, so their pens are cleaned out every second day.

I put the mechanical bucket on the front loader of the

tractor, gather my grape fork and shovels and make my way to the cow pens. Sometimes when I am cleaning the houses I listen to the radio or to podcasts, and sometimes I lose myself in the task. It is odd to say, but in cleaning out the cow dung I have often thought of the Zen monks and their sand gardens, raking and re-raking, cleaning, purifying, being immersed in their actions and thereby entering a meditative state. I think you can find meditation in the shovelling of shit, too. When the tractor bucket is full I drive up to the *dunkel*, which is a vast pile of manure and used straw, and empty my load. The *dunkel* will be spread on the fields in spring as fertilizer. Everything on a farm has a purpose and a future use, every action is part of a cycle: the dung of winter will bring the grass of summer.

When all the pens are clean I bed them with fresh straw. The cows seem happier. They rub their heads in the clean, dry bedding, giving themselves a sort of dust bath, like the buffalo do in those BBC nature documentaries. I smile and am glad of my work. Like any animal, the cow does not want to live in filth.

After an hour or more my cleaning is done and the jobs are nearly over for the morning. Now I can eat my eggs. Today they will be poached.

Ancestors

To speak of cattle is to speak of man, for cows have been our companions for nearly 10,500 years. The origin of the domestic cow has been traced by geneticists back to a single herd of wild oxen in Iran. That breed of ox, the auroch or ure, is now extinct, but it must have been something majestic to behold. It is these beasts who grace the cavemen's paintings in Lascaux and in the even earlier Chauvet Cave in France.

Cave paintings at Lascaux, France

Standing at over seven feet tall, much larger than ancient man and modern cattle, these giants must have seemed otherworldly to our forebears. Perhaps they were praised as gods, feared as devils. Strong and brave, there was no other animal like them.

The earliest auroch remains have been dated from 2 million years ago in India, but the auroch was itself descended from the *Bos acutifrons*, the sire of all bovines. With the cooling of the climate in the Pliocene period, grassland increased, which led to the evolution of large herbivores by the time of the Ice Age. These were the days of the megafauna of the mammal kingdom, when woolly rhinoceroses, giant mammoths, sabre-toothed cats and cave bears ruled. The animals in J. K. Rowling's *Fantastic Beasts and Where to Find Them* are products of her imagination, but all along nature has provided creatures just as strange. The auroch was the success story of the *Bos* family and it spread from India, migrating west and east.

Aurochs reached Europe some 270,000 years ago and so we could say that Europe belonged to them before it belonged to us. This continent was their great pasture, its ancient forests provided them with shelter and witnessed the cycle of their birth, life and death.

The auroch looked very different from our modern cows, perhaps more like a cousin to the American bison – with slender legs, an athletic torso and muscular shoulders and neck. Its horns were angled forward, and could grow up to a foot long.

As I child, I remember reading about the auroch in our worn and broken encyclopaedia and feeling a great sadness at its absence from my life. I told my parents of these magical beasts and they listened with great attentiveness. I think perhaps that night I walked out to the yard and looked somewhat miserably at our own cows and wished them just a little bigger.

I did not know it then, but I was not the first to be fascinated by the creatures. In antiquity, aurochs were held in awe, and even mighty Caesar praised their power and ferocity in Book 6 of his *Commentaries on the Gallic Wars*:

> These [ure] are a little below the elephant in size and of the appearance, colour and shape of a bull. Their strength and speed are extraordinary; they spare neither man nor wild beast which they have espied. These the Germans take with much pains in pits and kill them. The young men harden themselves with this exercise, and practice themselves in this kind of hunting, and those who have slain the greatest number of them, having produced the horns in public, to serve as evidence, receive great praise. But not even when taken very young can they be rendered familiar to men and tamed. The size, shape and appearance of their horns differ much from the horns of our oxen. These they anxiously seek after, and bind at the tips with silver, and use as cups at their most sumptuous entertainments.'

Maybe it was Caesar, or some of his generals, but the Romans brought some of the animals from Gaul back to their games, where men fought them to the death. It is not recorded who won.

Their great horns were prized as drinking cups by the nobility, which surely hastened the aurochs' decline. Indeed it is worth noting that the drinking horn at Corpus Christi College, Cambridge, is believed to be that of an auroch, given to the college in 1352.

The hunting of aurochs began in earnest then and over the centuries their numbers slowly dwindled, until hunting them became an exclusive right of the nobility and poaching them was punishable by death.

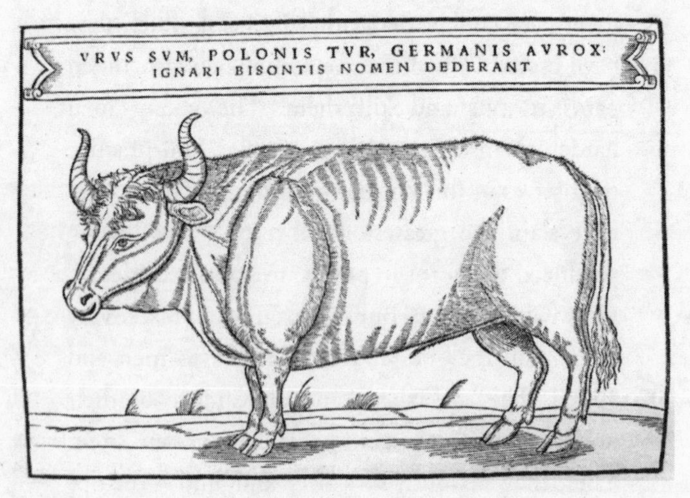

Auroch by Sigmund von Herberstein

This is the last known image we have of the auroch, produced in 1556 by Sigmund von Herberstein, a diplomat from Carniola, which is now in Slovenia. The beast, like the diplomat's own country, would soon cease to exist.

The last of the great aurochs died a natural death in Jaktorów Forest in Poland in 1627. The king himself had refused to hunt her and I imagine that he must have been sad at the news of her death, just as I was centuries later. There is a plaque to her there now. The last of her kind.

Calves

Every week I put down bedding for the calves in the area we've put aside for them in the big shed. This space, the creep, takes three square bales of straw, and is enclosed by gates, with a passage out to the slatted shed where the calves can go and see their mothers and suckle.

The creep is theirs alone. It is a safe place. In the late nights when I walk out to check the sheep, I see the calves huddled together fast asleep – they lie together like great dogs or fawns, curled and warm.

The bedding of the calves gives me a great satisfaction, for I know that they are clean and that I can keep infection away. They seem to enjoy it too, for they puck and jump and leap, sometimes butting one another in mock fights, instinctively practising for the freedom of the great meadows of summer.

The straw bales have sat in the shed for two winters. They are not heavy, but as I carry them across my back, one by one, I am remembering the sunshine that made them, and the man who baled them. Richard Monaghan was his name, a tillage farmer and agricultural contractor, he has been gone two years now. Sometimes I think of my memories of him when I carry those bales, of the harvest trips to his farm, the

September days that he and we had enjoyed. Changed now forever.

When the cancer came, it was quick. Of his illness I do not know so much, but he travelled to Dublin to the big hospitals, took his treatment, battled and did what the doctors told him. In the end, looking for relief or hope or thanks, he made a pilgrimage to Lourdes, where the Virgin appeared. Granny told me that he had been riding a bicycle there, like a young man, and then the day after he came home he died. It is said that the dying often get better just before the end so that they can say goodbye. Perhaps those few days were his blessing.

The bales he made are nearly gone now and his fields are worked by other men. Next season we will have to buy our straw elsewhere.

'Them calves need a good foot of straw – drown them in bedding,' Mam says as we meet at midday for coffee. She is taking a break from the school and I a rest from the farm.

'I've just done it,' I say.

'I read it in the *Farmer's Journal*,' she says.

I have seen the article too and we both agree that the *Journal* often has good advice. I know that it pleases Mam the same as me to have the cows clean.

'We don't want a repeat of that joint-ill,' I say.

'That was a terrible year.'

Many winters ago, when the big shed was still new and I was still in school, calf after calf succumbed to that malignant illness. We bedded and cleaned and disinfected, but the sickness seemed to have gained a foothold.

The bacteria enter at the navel and get into the bloodstream, where the disease spreads and attacks the joints of the animal. In the worst cases it can leave a beast a cripple. They do not need to be put down but they will never thrive and gain weight as they should. To see them walk with the illness makes you ashamed that you hadn't acted sooner, but these outbreaks do happen and one must be ready for them.

'I've got one or two of them with a slight scour,' I confess. It is a type of infection which leads the calf to pass dung excessively, leaving them dehydrated, and will, if untreated, kill them eventually.

'Well, that's not so bad. We have the medicine, don't we?' My father asks.

I nod.

'What with all the lambing, I missed them,' I say.

We finish our coffee and agree that it is time to return to work.

I meet Dad in the yard and tell him of the scoured calf.

'The French stuff, give them the French stuff.'

'And some of the pink mixed in?' I ask.

'Aye,' he says.

We have a small kitchenette in the sheep shed, where we keep the immediate medicine. It is a small apothecary full of powders and solutions, needles and gels. There is everything here for an emergency and a delivery. Hanging from a nail is my lucky strand of bailing twine. I have delivered fifty lambs with it now and it has saved many lives. Above the workshop hangs a woodblock painting of Saint Francis – it came all the way from Santa Fe, New Mexico. It is all I have left of a former life and love in Canada. Saint Francis has watched over us this winter, and sometimes in the night, in the blackness of it when I have delivered a lamb, I bless myself in front of him. That is our secret, just me and him. Saint Francis is patron saint and guardian of all animals. It was said that when he died the donkey which had carried him through life wept.

Hidden between the oxytocin and vitamin B, I find the French stuff. It is a green liquid with a French sounding name and I syringe out the correct dosage. I squirt it into a clean empty beer bottle and rummage through the medicines and extract the pink stuff. We have no name for it, for it was given to us by the vet, but it stops the scour and together the two substances kill the bacteria.

I shake and mix the liquids and wait as the kettle boils. On the radio they are talking of Syria and I hear the reports of the latest arrivals on the Greek islands. Ireland is in Europe too, but this feels so far away, so far removed. I cannot imagine the mass of humanity surging and swelling, for it is so quiet here and I see so few people now. There was a time when I worked in human rights, with people like this. I think of the Tamil refugees I met and worked with years ago as a journalist, and wonder where they are now. My work with others at that time led to their being freed from detention centres around the world, and I wonder now if they are happy in their new lives. Will these Syrians find new lives? I do not know.

The kettle boils and wakes me from my thoughts. I mix the last of the mixture and together Da and I seize the calf and shove the bottle down his throat.

'He's not so bad,' Da says as we finish.

'He had a bad dung out of him this morning.'

'Too much milk.'

'I don't think so.'

'That will do him good anyway.'

'It will,' I agree.

We do not talk much, Da and I, except about the sheep. That is our common ground, the ground we feel safe in, or he feels safe in. I am writing now, but we never talk of art or literature, nor when I was a journalist or a film producer did we talk of my investigations or film work. Only in the sheep can we truly communicate, and so the arcane language of breeds and lambs and ewes could be taken to mean, How are you, son? How are you, father? I love you, son. I love you, father. We have a world of mutual understanding and so long as we do not fight, we can both live in that world.

The released calf returns to its mother and we think no more of him. He is one of the oldest of this year's crop, and is strong and should recover quickly.

'I'll make the dinner,' I say and wipe the calf shit from my trousers.

'That'll do.'

Each day I peel the potatoes and vegetables and set the table. Mam says she can do it, but now that I am home on the farm, I want to ease her workload.

Today we have beef burgers and onions and soon the kitchen is steamy and hot. Mam enters at one, takes a cup of tea and waits while I finish the dinner. She calls Da on her phone and together we three eat. We talk of the day, of the Montessori school and of the animals. The weather is still bad but we try not to discuss it, for it is Ireland and the weather is always bad in winter.

Names

We do not name our cows, or not with Christian names anyway. They are given names that reflect their traits or the place where they have come from. Every cow has a story, just like every man. As I walk through the sheds in the afternoon, checking their bones to see who is near calving, I think of their histories.

There is the black whitehead who is an old lady, the dame of the farm, for she has never had a bad calf. She is quiet and calm and when I have to milk her she has never kicked me. I rub her flank and feel that she will carry another week or more. Walking along the bay, I move to the Simmental, who looks exactly like Daisy, her mother. Daisy went mad in the end and we sent her to the slaughterhouse for fear she would gore one of us to death. Her daughter is simply called the Simmental, after her breed. She is a docile creature, but for a week after she calves the sickness takes her and neither man nor beast can go near her. I have nearly lost my life with her more than once. Vinny has learned to stay away.

The sickness comes on some cows after they have calved. It is their way of protecting their young and they are simply best left alone. Their calves are usually fine unaided and so we do as the cows bid. Thankfully the Simmental has not yet calved.

I run my hands over the others, the Limousins and Blacks, and they lazily flick their tails at me, shooing me away. Vinny barks from the shed passage, for he is afraid I shall get hurt. I tell him that he is a good boy and that I am fine and he stops talking.

There are many cows whose stories I do not know, and nor does Mam, for Da does the buying. He has gone to marts all over the midlands, from the plains of Westmeath to the hills of Leitrim, where he used to meet the writer John McGahern in Mohill mart. They talked once or twice, he told me, not of books but of cows. I missed those meetings and miss them now, for there is so much I should like to have asked McGahern. He is the one modern writer my parents know, and many years ago we all of us watched *Amongst Women* on the television. We saw much of our own lives reflected in the character of Moran and his farming life. In Moran's quick fits of temper, we saw Da, and in the closeness of family, we saw ourselves.

McGahern knew farming, as do I, and at times when the work is good and the weather not so hard I think perhaps this life of writing and livestock could work.

I run my hand on the last Limousin and see that she is close to calving. I will move her down to the byre. She is old but you could not tell, for her coat is sleek and shining. I do not know where she came from or what farm she was born on and I must remember to ask Da.

The Limousin is a French breed of cow, initially used as

a draft animal. They remained largely unknown outside of France until the nineteenth century, but it is now one of the most popular breeds in Ireland, the backbone breed of many farms, for it has a good ratio of milk to meat. Limousin calves are smaller, but birthing difficulties are a lot less. They are a great cow, except for their temper, which is known by all. I have seen them break through ditches like racehorses, scale walls and fight bulls. They say red is the colour of passion; it is also the colour of the Limousin.

I move the newborn calf and mother from a few days ago out of the calving pen and make the house ready for the Limousin. This pen is our largest and has a calving gate, which my brother bought Da two years ago. It is a modern, galvanized contraption which locks the cow's head in place during labour. It makes calving a safer task for both man and beast.

The Limousin and I walk together through the shed. Her elder, which is our word for udder, dangles back and forth, engorged and full of milk. We call this being sprung. Her vagina is swelled and broken down and a thin clear slime hangs from it. She will calve in a day or two. She cannot tell me and so I must read these signs and understand.

I prod her flank and hush her towards the house. She lets out a low bellow, looks around her and enters. I give her silage and a small bucket of nuts, which have been manufactured with a smell that the cows love, and she quickly devours them. She will hold for now and I enter her name in the roll call of jobs in my mind.

Running

I started running a year ago. Sometimes when I am running hard I forget where I am and what distance I have covered. I forget the pain in my legs or ankles. I simply am. Exercise has become a big part of my life and most days I head to the local forest or the gym and spend an hour or two working out. It lifts my mood and gives me a break from the farm. To spend too much time on a farm surrounded only by animals makes an oddity of a man.

I've been a serious runner for the last nine months. By serious, I mean in my own terms, that twenty-one kilometres isn't hard or out of the ordinary any more. It started as a thought, a small want for change, in what the writer Yukio Mishima called the cultivation of the garden of the self. I was in Australia, my old home, promoting my first novel; I had emerged from a serious illness and the idea of fitness came to me suddenly.

At the beginning I was slow and would soon run out of breath, but as the weeks moved by my time on the treadmill increased, my heart grew stronger, and when I moved back to Ireland I graduated to the great outdoors.

I remember now my first five-kilometre race here last August. It was a simple course through the local forest and my only goal then was to finish. I had never before run so far and, slow as I was, I made it to the end. As I neared the finish I found I had more fuel in the tank and sprinted to the line. For the first time, I enjoyed a runner's high, a thing I had heard about.

Long-distance running is like farming, for it requires discipline, patience and preparation. You can't simply decide you will run a marathon on a Wednesday, just as you can't produce a herd in a day or make a cow calve before her time.

Come rain or shine or tiredness, I have been running here for twenty minutes every day (the twenties, I call them), to keep me fit, with a longer run thrown in once a week, to build my distances. There are days when I do not want to run, as there are days I do not want to farm. I think then of the Finnish long-distance runner Paavo Nurmi, who said, 'Mind is everything, muscle mere pieces of rubber. All that I am, I am because of my mind.'

I think about a lot of things when I'm out running. I think of the cows and their upcoming births. I think of their calves and the small ailments they have. I think of the fields, which are still hopelessly wet, and of the remaining silage and I wonder if we will have enough. I see them there in my mind, all these creatures dependent on me, on the seasons, on time. As I break my jog and enter a run, I push through the pain and pass the boy I was a year ago, who was struggling with so many things. I do not feel pity for him, just love, for I feel that I am in control now – my life has a sense of stability and order – and that I, like the animals, am safe.

There is a philosopher of running that I discovered, Dr George Sheehan, a medical physician and runner. His works have allowed me to understand running, and in a way to understand the farm, too. Man is an animal, he says, and it is our instinct to run, to maximise our fitness as animals. With every stride now I feel the strength coursing through my legs and I near the finish of my route; I think of that instinct for motion, of my connectedness with the cows, with our horses, with Vinny, with all animals that move and want to move.

Later, when I go out to the yard to feed the cows they know nothing of my run, but perhaps they recognize in me another beast, just for this evening.

Evenings

I like to be home from the gym or forest for half past four. The animals are hungry by then and the evening feeding must take place. Da will have been out in the yard for a while and sometimes we work together on the evening shift, but sometimes I text him on my way back from the gym and tell him to take a rest and I will do it, for I know the jobs I have left for myself from the morning.

First I fill two buckets of nuts and feed the weanling calves. They are growing fat now and their flesh will be marbling nicely. The bullocks and heifers are separated and bellow at the sound of the silo latch opening. The nuts spill out, rattling to the waiting plastic. The weanlings get beef nuts twice a day and as I pour them into their trough the animals puck and jostle for position at the feeding barrier. Occasionally a weanling will not like nuts, so they will not grow as quickly as the others, but we cannot change that.

The weather is so bad that I wear my fleece at all times

on the farm, and it now smells of amniotic fluid from all the lambs I have delivered. The weanlings do not know what to make of me, for I look like a man but smell like a sheep. I know I have confused them on many occasions.

Next I must feed the mature cows, and for this I will need the tractor. It is a John Deere, which is, I suppose, the BMW of the tractor world – reliable, strong and popular. Ours came from Northern Ireland and is getting on now. Its starter has been giving us trouble these last few weeks and it is a battle to get the motor going each day. I turn the ignition but the motor does not spark and I must take my wrench to the alternator once more. I do not know what this does, but I have watched Da do it many times, so I imitate his action and after several minutes of trying, the motor clicks and comes to moving life.

I prepare the round bales of silage and pick up the bale with the tractor's front loader, a mechanical arm with a large blade attached, then step out of the tractor to cut the plastic from around the bale. The mature cows will eat two bales. I drive to the automatic feeder and I unroll the plastic netting which keeps the fermented grass in place and load the bale. The feeder will roll and chew the grass and spit out a neat row of silage for the animals. All I need to do is drive along the passage way and direct the flow of food. As soon as the tractor enters, the cows wake from their slumbers and call to one another in excitement.

To watch this display is to know that these animals have cognition: they know the tractor and that it means food. They have memory and thought and, while these faculties might not be the same as those of man, they are aware of this world. I have read that cattle can remember human faces for up to a year. To think in this way I wonder if they mind being cooped up in here. But there is so little in the fields this time of year and the sheds are warm, so perhaps they are as happy to be inside as we are.

On account of the bad weather, we have kept the bull inside this year, too. He has made his presence known in the slatted shed and has at times pucked calves and mounted in-heat cows. When he does so, we must take note and see whether the cow holds and goes in calf. We have separated the pregnant cows from him, for he might try to rise on them too, and in so doing hurt them or cause an abortion.

I load up another bale and drive to the lower shed to feed the weanling calves. They have greedily finished the nuts I poured for them earlier and are waiting for me. We keep food in front of them at all times and they gorge throughout the day.

Lastly, I give the small remaining silage to the sheep. Their bleating is incessant and does not stop until each and every one of them has been fed.

In comparison to the cow, the sheep is a gentle but stupid creature. I have devoted many hours to their care, but I do not think they know who I am. I get the sense that every day is a new day for them, that they are surprised anew by my appearance each morning. I'm told sheep can remember human faces, but I've seen no evidence of this yet.

I check their water and make my rounds to the pens for the newly lambed mothers. I give them each a small bucket of nuts and hay. There are some lambs which need extra feeding and this takes an hour or more, for I must heat up milk and bottle-feed them. I cannot rush this process, for the lamb will only drink at one speed. Sometimes I kiss their little foreheads, for they are so gentle and innocent. I listen to the radio and sit and think.

The radio is always on in the sheep shed. Da said he read in the paper that the radio gets the lambs used to the human voice, so that they are not spooked when we come out and talk. So far it seems to have worked. It is company in the night for us too and there have been times I have heard beautiful music echo through the darkness. I do not know if sheep like Seán Ó Riada or David Bowie, but I have heard both out here now and they didn't seem to mind.

This evening the chords of a Johnny Cash piece creep through the shed and I smile in recognition. It is a reminder of an older life, a different life. I see myself in recital halls and music venues, talking over coffee and sipping on beer. I see too the money wasted in cities, pissed against walls, and the oppressive encroachment of urbanity that always in the end forced me back to the countryside. In the past I have seen myself as the courageous journalist, as the wily, tight-fisted film producer. Now I see myself as the farmer's son I am, and think that all the rest has just been an act. The country mouse playing the city boy for a time. I know the laws of the fields, the ways of land and cows, but the code of cities, the laws of film or media-industry types, is different; I am not trained in such rules and it took nearly a decade for me to learn this.

But it was not all bad, there were times of great happiness in those urban environments, in those farms of men. That is where I met Vivian, my girlfriend, years ago and thousands of miles away in Sydney. We were both very young and so many other things have happened since then.

The 6 p.m. Angelus begins to chime on the radio and wakes me from my dreams. The lamb has finished his bottle, his brother too must now be fed, and I repeat the ritual. Heating the milk, testing it upon my forearm and nursing him in my arms.

I sit upon a small milking stool borrowed from an old neighbour's house. I am in no rush and enjoy the process. The bells ring out and there is silence in the shed now, for all 150 animals are eating. Vinny patiently waits in the alleyway for me. I have not had time to walk him today, but I must remember tomorrow. I must not forget his training. I treat him now with a rasher each day after our walk and work. He is learning fast. Even Da has praised him.

On the way back to the house, I look in on the calves playing in their creep. The sick one I gave the French stuff to is asleep. I check his nose and it is warm and I feel that he will recover. He opens his eyes slowly, recognizes me and springs to life. He shits as he jumps and the dung is still runny but not as much as before. Its colour is returning to normal and I smile, for at times it is like the movie *The Madness of King George*, and I am the physician inspecting stool samples.

'You'll be OK,' I say, the first words I have spoken in two hours.

Finally, I check the Limousin. She is pacing her house, rooting up her straw. She will calve tonight.

It is pitch black now and the lights of the sheds guide my way back to the house. I begin and end the days in darkness. The lack of sun has made itself known and the moods of the parish are low. We all think of the sunshine to come, and wait.

'It's bedtime, Vinny.'

He barks and runs towards his cage. I check his water, give him food for the night and close the door. We have bedded an old diesel tank for him. It is warm and dry and he cannot wander in the night. He is safe and so are the lambs.

I am tired. It has been another long day.

Gods

With the domestication of the auroch, two subspecies appeared: in India, the Zebu, and in Europe, the Taurine. All modern cows belong to either of these subspecies, and both have a long association with divinity.

The Indian Zebu is thought to have emerged during the Bronze Age, and was characterized by a distinctive hump and large dewlap, or flap of skin, below its neck. Within the pantheon of Hindu gods, the animal was represented as Nandi, the sacred bull and *vahana* of Shiva. The *vahana* is the carrier of a god, much like Saint Francis's donkey.

Shiva chose Nandi as his *vahana* because the people of

India at that time were mostly farmers and cows were the main form of transport. Shiva was not a warring god and chose to spend his time meditating and in thought, so it made sense that he would be carried by a slow but reliable animal. Nandi had strength and calmness and virility too.

There are temples and statues to Nandi throughout the Indian subcontinent, and in some parts of India people still worship at his shrines. It is said that if you whisper your wish in his ear it shall be granted. In Sanskrit Nandi means 'happy' and in old Tamil it means 'bull', so we could say that he is the happy bull. Happy indeed, for the cow is a protected and sacred animal in India.

Even older than the Zebu, the Taurine emerged in the Near East some 10,000 years ago. This was a period during which settled farming and agricultural technology spread from the fertile crescent of Mesopotamia to the first great civilization of Egypt. The cow was part of this transition, working the fields, giving milk, meat and even warmth on cold nights, so it is no wonder that the animal was held in high regard. And, just as Nandi acquired a god-like status in India, so in the era before the pharaohs the Ancient Egyptians worshipped Apis the bull.

Judging by the later hieroglyphics and statues, Apis was from the Taurine species of cow, for he is without the distinctive hump and dewlap of the Zebu. He stands proud, muscular and strong as a modern European bull.

The most important of all sacred animals in Egypt, Apis was associated with strength and fertility in relation to agriculture. He was also the servant and manifestation of the original creator god, Ptah, who thought the world into existence and gave life to all things through his words. We cannot be sure, for not all the records exist, but it seems that the bull might also have been the representative of a king who has become a deity after death. Perhaps it was Apis that the Jews worshipped as the Golden Calf when Moses went to the mount. For he too was the sacred cow, the golden calf, the bringer of fertility. Indeed the Canaanites, another Semitic people, worshiped their creator god, El, in the form of a bull called Toru El.

And it is in death that Apis, the sacred bull, would be given his greatest honour, for the Ancient Egyptians celebrated the afterlife in all its glory, pomp and sadness.

The *Apis papyrus* records much of the ritual of the mummification and burial of the bull. The instructions were so

Egyptian relief with a bull and an ankh, the symbol of life, from Luxor Temple (Thebes).

precise that one might suppose that those carrying out the ceremony feared that any deviation might prevent Apis from returning to the heavens and so prevent his rebirth, thereby upsetting the very balance of life itself.

The process took seventy days and the priests involved could not bathe during that time. They were expected to wail and mourn throughout their work and to adhere to a strict diet and fasting, which included not eating milk and meat. I imagine now that I can see the ritual once more. I can smell burning sage in the air and hear the mournful chants rising at the loss of Apis, as he is laid out, god-like, upon the mighty stone table. He will be washed and cleaned and then mummified. His head and mouth will be embalmed first, and then the body, once the organs have been removed. His cows will cry out for him, as I have seen our cows do at the loss of a calf or the taking away of a herd member, as he departs on his final journey.

The mummified bull was transported to Saqqara, the city of the dead near Memphis in Lower Egypt. It was here, some 5,000 years ago, that the oldest-known complete building complex in history was built. These *mastabas*, or 'houses for eternity', were flat-roofed structures made out of mud bricks with sloping sides, and were the precursor to the pyramids. Inside, the bull would be placed inside a colossal seventy-tonne black sarcophagus. It was here that the bull would make his journey to the heavens.

The story of Apis does not end there, however, for upon his passing the search for his reincarnated self began anew through the herds of the country. This must have been akin to the search for his Holiness the Dalai Lama, for the records are very clear that certain markings and traits must be on the chosen beast. When found, the bull would be raised by hand, given a harem of cows and would live a peaceful existence, tended by priests and minders.

In 1850, the Serapuem of Saqqara was rediscovered near the Pyramid of Djoser by the French scholar Auguste Mariette. All of the tombs in the temple had been robbed bar one, which is now in the Agricultural Museum of Cairo, but there was a large collection of some twenty-five Apis bulls, suggesting that the practice continued through the successive dynasties. Such was Apis's importance that Alexander the Great himself made a sacrifice to him during his recapture of Egypt from the Persians.

The cow, the wild creature, had not only been domesticated, but had also entered the spiritual world of man.

Equus

We have always kept horses on the farm, not for riding or status but rather, I think, because of the traditional Irish connection to that species. We are people of the land and the horse has been our way of life far longer than the tractor. In Mam and Dad's childhood, every farm had a horse or donkey to perform the work, pull the trap and break the soil.

The nearby village of Ballinalee has a Connemara pony show each June, and I think it was perhaps here that Da got the idea of having a breeding mare. Ashling was our first horse and each year for many years she bred us foal after foal. We always got a good price for them and she earned a place of respect on the farm. Ashling was just a few days apart in age from my sister Linda, and, as I remember it now, we often celebrated both birthdays together.

The Connemara is a native Irish breed from the west of Ireland. They say the breed has Viking blood, and that when the Spanish ran aground off the west coast their Andalusians broke loose and mixed with the native horses. Whatever the blood, they are the horse of this country. They are small but graceful and make for great show ponies. Ashling's sister had competed at the Dublin Horse Show, one of the great world showjumping events, and I heard she was later sold to a man in France. I never knew her name.

During the years of the Celtic Tiger, which was our economic boom from 1997 to 2007, there was a great demand for horses amongst the rural people and we expanded our herd to ten, but then, with the collapse of the economy, no one wanted horses any more and we sold many of them cheap. It pained my father the day we took them to the mart, for they were all of them great ponies.

I had named one of them Grey of Macha, after the famous stallion Liath Macha, the chariot horse of the central hero of Irish mythology, Cúchulainn. It was said that he had emerged from a lake as a gift given by the gods to the warrior. The Cúchulainn story takes place in these lands and I imagined that perhaps that horse once galloped across these very fields. Grey of Macha was sold as a stallion, wild and untamed, but he would make a good breeder for the right man. The others went to the continent and their fate was not so blessed, for they were bound for an abattoir.

Before Ashling stopped breeding, she gave birth to her last and greatest foal, Hazel. Hazel was praised by all the callers to the house and many men came from far and wanted to buy her, but Da always refused. She would replace her mother, he said. It was with great pride that my parents accepted the blue ribbon for best foal that year at the Connemara pony show and then they retired from the shows and bred no more.

Ashling died peacefully of old age in the front field years later, but we still have Hazel, as well as our donkey, Asal (pronounced *Aw sal*), which is Irish for 'ass'. They are outside in the boggy ground and pickings are scarce in winter, so I feed them a square bale of hay each day. A few weeks back, on the one day I had not fed them, the donkey's foal died of some unknown illness. I have thought many times of the ifs of that situation: if I had gone out that day, perhaps I could have saved him; if I had gone out early the next morning, perhaps he might still have been alive; if, if, if . . . But we cannot bring back the dead.

And so Asal and Hazel wait for me now each morning to come with their hay. It has the smell of summer and the feel of sunshine. I know they too can taste it. They run to greet me and I hear their neighs and brays of excitement. I open the bale and let them eat; some days I rub their flanks and speak with them. They are social creatures and need to be spoken to, lest wildness set in again.

This hay is three years old, for there has not been a good

summer since to make new bales. The hay is from my uncle Mick's ground. He too is dead. We have rented part of his farm from my aunt, and I think it would make him happy to see his fields used again. It is strange, for as Mick grew sick with the lung cancer so too the weeds began to choke the land. It was as if a sickness had come upon everything.

They are the fields of Da's childhood, for he worked them with his own father as I work them now with him. To have these fields in his possession, if only rented, is an important thing for him. We have made them healthy again and, when the weather is dry, green grass grows. Da has never spoken of that fulfilment of the land or the fields being green once more, but I know, I feel it; Mam has felt it too. The hay, like the straw, like the cows, has memories of other times.

Uncle Mick was gifted with his hands. Everyone says he should have been a mechanic or engineer, for he was happiest in engine grease and oil. I think back to the summer of childhood when we came down Soran Hill on a three-wheeled tractor (the front left wheel had got a flat and we had no spare) with a full load of square bales on the trailer. I sat atop it all with Mick's son, my cousin Michael, and we had never been so high in our lives. As the three-wheeled tractor drove down Soran Hill and took the turn for home and the hayshed, I remember us both holding on for dear life. Michael is grown now, with a child of his own. Mick's tractors are sold, his land divided and changed. We talk of him still, as though in the same breath he might just appear in the yard once more, with his peaky hat and the butt of a cigarette in his mouth, but as time moves on the talk becomes less.

The death of Mick fell hard on my father, for they were more than brothers – they were friends. I think now as I rub the horses, watching them eat the last of the feed, that that loss has hardened his heart in some way. He does not go to mass much any more, and perhaps he is angry with God at the loss of Mick and his other brother John, who died suddenly three years ago.

Uncle John planted the birch trees around the farm, the trees that have given their name to this place. In them I see him and his quiet labours. His heart attack broke us all in new ways, and it still seems all too fresh to talk of.

And so each morning I carry my bale of hay from the shed, and in some respect I carry the total memories of family and men who are no longer here. I walk out to the horse and cross the gate, throwing it on the ground.

Hazel neighs and I pet her before heading back to the farmyard. As long as we are here, there will be horses on this land.

FEBRUARY

The Way of the Cross

It is the first day of February and I am out in the fields cutting rush, a type of reed which is common to this land. The paddocks around the house are reclaimed bog, and there is a fine crop of this weed. I cut and gather and bundle, and soon I have enough to complete the job. It is the feast day of Saint Brigid, and I am collecting the blades to weave a cross of rush, as it is said Brigid herself did when she explained the way of Christ to a local Celtic chieftain on his deathbed. I do this because we have always done this. I do this in a way beyond religion. I do this in a way of culture.

Mam has asked the children of the playschool to bring in their own rushes and she will show them how to weave the cross so they can each of them take one home.

As a youngster growing up, I often wished that we were French or Spanish, for their culture seemed so vibrant, so alive to me in its tongue and rituals. Our own seemed so watered down, so globalized, that I had wished myself someone else. And yet, in these last few years and after these last few months of farming, I now see that our lives are rich with tradition.

Brigid is celebrated on 1 February, but it was once a Celtic feast day called Imbolc, which marked the beginning of spring. Back then, another Brigid was celebrated that day: the Gaelic dawn goddess, the daughter of Dagda, the good god. It was said that on Imbolc eve, Brigid would visit virtuous households and bless the inhabitants. The practices that I and my family still carry out are ageless; we continue to live out our Celtic past. In my cutting and gathering, I am worshipping both the old and the new. As so often in this place, everything is done in an evocation of something else, something older.

We were once a tribal people. My mother's family originally came from the race of Bréifne in Cavan, home of the O'Reillys, and my father's from the Mumu in Kerry, home of the O'Connells. The language of our ancestors is long changed but it and our culture still flow in our veins, and you have but to look into our faces to see it. Tribalism is not dead. We have an attachment to this land that goes beyond money; it is a connection of a spiritual quality. It is our *baile agus beatha*, the place where we come home, that which sustains us.

In my years as an emigrant, I have met only the Aboriginals of Australia and the native peoples of Canada who have fully understood this connection to place.

When I was twenty-one and finishing my journalism studies in Australia, I made a documentary about the Aboriginals for my thesis. For a number of months I lived in the Northern Territory and shared my life with these people of the desert. I found in them a reflection of what I knew from home. Their struggles for land rights and native title did not need to be explained to me, for I understood that it was their *beatha*, their spiritual link. To borrow another Irish word, it was their *draiocht*, that magical connection which is unseen and without which they miss a part of themselves and are powerless.

I remember the day I visited the town of Katherine, where an Aboriginal health worker told me of an old man who remembered his first encounter with the white man. They had come with cattle and guns in what is called the Frontier Wars. The old man had been a boy then and everyone in his tribe had been shot. He alone had survived. The health worker looked at me then with tears in his eyes.

'We lost so much,' he said.

I nodded. The sun shone down on us and I did not speak for a time.

'Who was the old man?' I asked then.

'He was my father.'

Freedom is a wonderful thing and we are lucky to have it. But we are both peoples of colonialism: the Aboriginal people can never regain all that was taken from them and neither shall we. Perhaps this is why our culture endures so strongly, why the rituals of the past take on new meaning and significance in this modern world. I work the rush as the Aborigines do their timber for a throwing spear, or as the Northern tribes do for their didgeridoos.

I take the reed in my hands, now bend and fold and tie its four ends with string. I will place it above our icon of the Virgin and child which hangs over the fireplace. There it shall dry out in the days to come, ageing and changing as the year goes on. I replace the cross each year – it is a topping up of faith.

Now that I am home, I go to mass each Sunday. Our parish is Killoe and there is a healthy congregation. I go on my own and listen now more deeply to the scripture than I used to as a child. I find great beauty in its words and stories, and it has made me think. Its imagery is often that of the farm and farmer. The scripture is rich with men of my trade, men who have risked everything to look for a single missing lamb and felt happier at its finding than in the keeping of the other ninety-nine in their flock. I too have known that joy at the salvation of something you thought was lost – as shepherd, as cattle drover, as a man come through darkness.

Religion has become a laughable word in modern life. To believe is often to be scorned, and yet what is so wrong in belief? I do not agree with everything the Catholic Church says and there are things I ignore. I too have doubted my faith, even turned my heart from God, and yet I have found faith once more. I found the beauty and wonder of nature on this farm, and in it the joy and despair of life. For this, I have no other word but Yahweh.

Father Seán is our parish priest. He is also my *anam cara*, my soul friend. We meet most weeks and talk of literature until the late hours. He is the best-read man in the parish and the one to whom I show all my writings. I have not written in four months, for I have been so busy on the farm, but we talk of what the next book shall be, if it shall be. He believes in me, in my work, in my decision to be back in Ireland and write for a year. Above all he has faith that my writing will see the light of day.

On Sundays at mass we do not speak, for he is at work and

I think perhaps he cannot break character while he is at the altar, at the coalface. Sometimes I stand at the back of the church and watch the people of the parish and wonder about their lives. I do not know everyone here, but I do know that we are Father Seán's family; he is our shepherd and cares for us all. Like many of us, he is a tiller too, a tiller of the soul.

After mass is ended Father Seán and I bid one another farewell. He has returned to being a man once more and we can speak. I wish him well and tell him I will call over during the week. He alone knows that I once contemplated becoming a priest.

Nights

In the evenings as I walk across the yard to the house, the customers for mother's Montessori school are arriving to pick up their children and her day will be finished soon. I say hello to some of them and they return a smile or nod. They too are tired and many have come from Dublin offices some two hours' commute away.

I come inside, I take off my overalls, wash my hands and hang up my fleece.

I prepare Mam a sandwich for her supper each evening. She takes care of all of us but forgets to feed herself, and so I leave her with no excuse: the sandwich is waiting on the table for her each evening. Tonight it is ham and cheese with a nice salad dressing and some crisps on the side. She does not like fancy food, preferring simple and wholesome things, things of our culture. I will make her a mug of tea when she comes in the door. I fill the hot water bottles for the bedrooms, for the nights are cold now.

Da is going to the mart again with Davy tonight and they are selling a ewe that lost her lamb. She has not the kindness in her and would not take with any foster lambs, so we agreed that the mart was the best place for her. I leave a mug of tea for him too and a sandwich. Uncle Davy soon arrives in his white Land Rover and they head off, the ewe standing in the rear of the vehicle. Good riddance to her, I think; she was not meant for this place.

For the first time since lunch, I can now sit down. My little sister Javine has finished her homework and plays on her phone. She is the baby in the family, for my other siblings are grown with families of their own. Javine has entered her first year of secondary school. She is consumed with the social-media world and talks with all her friends through Snapchat and Viber. Witnessing this, I am finally starting to feel older and out of touch, but I do not mind this. Each generation has its own way of communicating.

The night shift starts around 7 p.m. and we walk out to the yard every two hours or so and check that the sheep are OK and that no new lambs have arrived. You cannot tell when a sheep may lamb, for they go into labour so quickly and, if all is well, can birth in twenty minutes. The cow is different: she gives us her signs and because a calf is big it takes many hours for her to go into active labour. I have been watching the Limousin and she me, and I know that she will calve tonight. I have the jack and ropes and masks ready and I have prepared a fresh batch of iodine and cleaned the stomach tube.

Da knows the cow will calve too, but he knows also that I can do it on my own now and, though we do not discuss it, I know he has handed over this small bit of control to me. I will be calmer tonight than I was with the red cow, for it is a simpler job. The Limousin is old and should calve without trouble. She is roomy, as we say, which means her passage is big enough for a man to fit both his hands in and wrestle a calf from her.

For now, though, I settle with my book. I am back reading after several months' absence and, from the world of births, cows and rain, I enter the jungle in *The Narrow Road to the Deep North* by Richard Flanagan, the Australian author. It is a long book about a group of Australian prisoners of war on the Thai–Burma death railway in the Second World War, and has kept my attention as I return to it night after night. At each chapter ending I break, check the time and see if I should go out to the yard.

There are nights when instinct tells me to leave the jungle and the struggles of those men. I cannot describe it any better than that – perhaps it is the sense of nature, or birth, but oftentimes on these trips to the yard I have found a ewe just about to give birth.

A few weeks ago, I had a battle of some half an hour to take a lamb from his mother. I was alone and it was raining and my lucky twine was failing me. The lamb's head was so large that I could not get my hand inside the ewe's passage to get his front legs, which need to come out first before the head. After much manoeuvring I had to place a rope around the back of his head (a risky move, for it could snap his neck or break his jaw), push the head back inside, and then fish his legs out.

The ewe was tight and I used large amounts of gel to lubricate my hands lest I should tear her and cause a haemorrhage, which could kill her. It had already been fifteen minutes since I had discovered her and time was against me. Normally a birth takes no more than five minutes, so panic was now setting in. I breathed deeply, asked the lamb to stay alive and dived in once more. My fingertips finally found his other missing hock and I pulled. His legs were slippy with fluid but slowly I felt his joints bend and come into place and a pair of hooves emerged before me. Taking another twine, I bound those feet in a double locking knot and then pushed them back inside.

I pulled my two ropes and his feet and head emerged correctly and he broke from his mother. He fell to the ground and gasped and I smiled and thanked God. Clearing the mucus from his snout, I took him up bodily and swung him in the air three times to clear any fluid from his lungs. After this was done I rested him back on the straw, his legs splayed out. His mother, now recovered, stood and turned to lick him. The bonding began and I collapsed onto the floor of the pen, the adrenaline draining from me. It had been one of the most difficult births this year, but the lamb – a ram – was magnificent.

Tonight, I check the sheep sheds but there is no new life or encroaching life. Vinny wakes in the darkness and barks out to me from his house. I tell him it is only me and to go back to sleep, which he does. I check the Limousin; a slime hangs from her passage as it has done this last few hours. She is not ready, not yet, I tell myself. She has been pacing in her house and her tracks have made a mess of the straw. I will put fresh bedding down when the calf is born. It will be another two hours at least.

It has been a long day and I am tired now. I usually go to bed at ten when I am on the night shift. Da will then make a last check at midnight and wake me if there is an emergency. I will sleep until 3 a.m. and waken with my alarm and check that everything is OK. If we did not have the sheep, we could sleep through the night, but during lambing season each night is like this. We are six weeks into the night season and I have become used to the routine now.

I lie in bed and sleep takes me soon enough. I have been dreaming of the sheep and cows lately. It is a sign of stress, a sign that the work is consuming me, but I cannot control my dreams, as I cannot control the workload. It is a lot for two men – two men who do not always agree. But we have not fought in a long time, so perhaps that is all in the past. In any case, he will be happy tomorrow because he had the night off; he will be happy for a few days.

At midnight I waken, for Da has not come home yet and I can hear the Limousin roaring, shouting with the pains of labour. I have slept in my jumper and socks and so quickly slip on my trousers and I am back in the yard. From her passage two feet hang and I know that it is time. I sense too that she has given up and that she will push no more.

I walk up the yard and fill a small bucket with nuts, the sound of which makes all the weanlings roar. Returning, now I coax the Limousin into the headlock. I place the nuts in front and as she walks in I quickly close the latch. Now my work can begin.

She does not fight or strain and I think she knows that I have come to help. I reach in and take the calf's feet and trace my hands up the legs to ensure that everything is coming correctly. When I feel its head, I sigh with relief, for I have not yet delivered a backwards calf.

The cow roars and her cry is answered by her sisters in the upper shed. If this birth were taking place in nature, perhaps they would surround her and keep her safe. I place the ropes on the feet – they are small feet, I notice, and think now perhaps he is a twin, for they are very small feet indeed, no bigger than those of a large dog.

'First one first,' I tell myself.

I nod and set to again and take the calf quickly. In the end I pull him from her with my hands. I carry him with one arm to the straw and he is so small and red and wet, but he is alive and that is all that counts.

Her vagina splutters and a breath of air leaves her and I think, now there must be another, a brother for the wee one, and so I check again. I run my arm inside her but, search as I might, I only feel placenta. It is mushy in my hands and I push it to one side, thinking perhaps his brother is behind it, but there is nothing. He is alone – and a dwarf.

I push the mask onto his face, but I know he does not

need it, for he is breathing fine. The Limousin moos and calls to me, and I know she wants to see her calf, but not yet, I must milk her first. For I wonder if this little scrap will even reach her tits.

As I milk her he makes to stand, as if to prove to me he shall live, and by the time I have filled my jug he has walked to me and is sniffing at her udder.

'Well, fuck ya, anyway,' I say, and smile, and it's all I can do but laugh at this, my little Napoleon.

I break the seal on her four tits and ensure the beestings is flowing and then, bending low, put the tit inside Napoleon's mouth and strug. The milk shoots out and within seconds he is sucking by himself. I help him stand for a time by placing my knee beneath him to keep him upright. This is better than any stomach tube, for he will learn the art of sucking straight away and will not need to be instructed as some calves do.

Napoleon drinks and drinks and drains one tit and moves to another.

It could be the Limousin's age that has caused the small calf, or a lack of vitamins and proteins. I am not sure. Perhaps she needed more feeding, and yet I look at her now and she is in good shape: she is rated as a five-star cow on the new government-introduced system. Perhaps the stars count for nothing, as I have heard men say. Napoleon may be her last calf; she may not be here next season.

I wash the blood from my hands and walk back to the house. Da has not returned yet, so perhaps he went for a pint. Mam shouts to me from her dark bedroom.

'Well?'

'Little bull,' I say, 'pipsqueak.'

'I'll look at him in the morning.'

'Blink and you might miss him.'

'Goodnight, John.

'*Oiche mhaith*, Ma,' I say, which is the old language for 'goodnight'.

Three a.m. will not be far away now and I return to sleep quickly, the smell of new life in my hair and clothes.

Horns

The weather has not improved. There was talk that February would be better, but then the snows came and that talk ended. Snow transforms the land, covering all with its sleety whiteness. There is a beauty in it but also a hardship, for our animals outside could get cold. The tractor is slow to start with the weather and I am wrapped up in several more layers. The snow has lasted for several mornings and the children of the area have made men from its whitey down.

Napoleon is alive and well and we have moved him to the big shed, and the weanlings are nearly ready for market, but the days have bled into each other of late. Da has not been out on the farm for a week, for he has had a bad cold and I have run the place myself. It has been a busy week and I am tired, for between the days and nights I am but a servant to the cows. Sometimes I have wondered what is it all for. I do not earn money at this work and the farm pays for itself and no more. To make a living at farming is hard work and there are few full-time farmers in the area; most men have other jobs as builders or tradesmen or teachers. Da is one of the few full-time farmers, but that was not always so. For more than two decades he was a builder with my uncle John, but he retired ten years ago, for the work had grown too hard and, though he was still young, it had aged him.

My brother now runs the building company and Dad the land. I wonder at times if it is enough for him. There is always work in the fields, yes, but there is no sociality to it, no spontaneity. As a builder, Da was a man of action, of business, and this life is so very different. The last few years have been hard for him, as he has lost friends and brothers. We never talk of that, but just once he confessed he dreamed of them. In the dream, he and his brother John were building a wall, as they had so often done in life. John was mixing the cement and he was laying the block; he said he could hear John's voice and smell the smoke from his pipe. They were happy, he said, and then he woke and could remember no more. He never spoke of that dream again.

Today is Da's first day back on the farm. The first batch of this year's crop of calves are getting older now and we must burn off their horns. Dehorning the cows is a requirement from the department of agriculture, for the meat factories will not take a horned beast.

We burn the calves' horns now to save sawing them off when they are older. We use a gas torch to heat a small metal bowl that cuts the horns from their heads. It is sore for the calf, but it is better done now than to wait, for to saw the horns off a weanling is a bloody job, for the horn must be cut at the root to stop further growth.

In my youth, I remember that we used this sculling method to dehorn the weanlings. Da had never the stomach for it, for he hates blood, and so Uncle Mick was our butcher and nurse. I still remember the cows' screams as Mick sawed through the bone and flesh.

Mick knew the job well and the beasts did not suffer for long. We would place them in the halter and pull their heads tight to the top bar of the crush, to prevent the animal from becoming injured during the operation. Once the hacksaw had cut the horns, the veins would shoot blood into the air and these had to be closed. Mick's own method for this was the use of tough rubber bands which tied and criss-crossed around both horn stubs and slowed and eventually stopped the blood flow. The bands were tight and required strength to put on. We would move quickly as the blood spat out onto our hands and faces, for if left unaided the weanling would bleed out. After fifteen or twenty minutes their bandaged wounds would begin to clot and the job was ended. The concrete of the yard would be littered with horns and plasma. I am glad we no longer farm this way. We have not Uncle Mick to do it anyway.

There are twelve calves to be treated today. We lock them in their pen and set up our calf crush, which is a small metal box. Da brings his equipment out and sets it up on the creep wall. He points out potential problems and I nod in agreement. It may have been a week since he has been on the yard, but it is still his yard and I think perhaps this is his way of letting me know that he is in charge.

'Grab that first calf,' he says.

I wrestle the first of the crop, taking him by the tail and ear and steer him towards the crush. I must keep close quarters, for though the calves are small, a kick in the right place could hurt me. By staying tight to the animal, he cannot lash out. I push and pull and steer and he is in the crush. As well as burning his horns, we must tag him and dose for blackleg.

We did not always dose for blackleg, but we learned our lesson years ago, when we found the prize red Limousin calf stretched out in the hill farm of Clonfin. He had been the biggest and brightest of that year's calves, but the illness had taken him. I can still remember the shame when the dead lorry had come to take his body away. Blackleg can survive in ground as a spore for years, lying dormant until conditions are right for it. No one knows why, but the bacteria strike at the best animals. The bodies of the dead must be immolated.

Da lowers the torch towards the calf's head. It cries for its mother as the hot metal pierces its flesh, he shits himself and squirms in the crush. I whisper soft words and tell him it will be over soon. I see my father scoop up the small piece of horn contained in the torch.

'Just one to go,' he says and bends to once more.

The calf cries out again and then it is over. We daub Vaseline over the open wounds to prevent infection. Then we tag him, giving him his number in the herd. Each animal born has a card, which is their passport. It records their life from birth to sale to slaughter – from farm to fork, as the slogan goes. Finally we give him 10 cc of blackleg vaccination from the small yellow box. We spray a small mark of paint on his back to let us know we have treated him, and he is free to go.

'Next one, Johnny.'

I can tell that Da is in a good mood now, for he only calls me Johnny when things are going well. I nod and grab another calf and we repeat the process. Some fight with me and I enjoy it all the more, for the challenge of man against beast makes me realize the strength I have built up from all my fitness training.

Sometimes I ask Da to tell me stories of the long ago. I have heard these stories many times but I do not tire of them. I think in a way it is these stories that have given me my gift of writing, as we call it here.

'Tell us that one about your uncle and the strong man.'

'Ah, you know that one.'

'Wasn't it the circus?' I ask.

'You know it was.'

'Well, tell it anyway,' I say.

'The circus came to the village, years ago now, and your granny's people, the Mullens, lived in Rhyne then. The circus had a strongman and the bet went, whoever could lift heavier than him would win five pound.'

'Which was a lot in them days,' I say.

'Be close to a thousand now, I suppose. Old Mullen lost the first year, but he got thick about it and decided he would win the next year when the circus came back. And so when the next calf was born on the farm he lifted him every day up over his head. So by the time the strongman came back a year later, Mullen was as strong as an ox and the calf as big as a bullock.'

'And he won the money?' I ask.

'He did,' says Da.

'Some lad.'

You would not think we had ever had in the parish our own legendary strongman, our own Milo of Croton, and yet Mullen of Rhyne existed.

The last calf escapes my arms and I hear Da cluck. He is growing tired.

'Ah, will ya catch him!' he says, irritated.

'I'm bloody trying,' I say, and I can hear now our tones change.

I chase the calf around the creep and finally capture him in a headlock. I stand to once more and direct him towards the shoot. I push him inside with my legs and we quickly bolt him into place. He is strong and his coat smooth.

'Is this the scoured calf?' he asks.

'Can't you see it is?'

Da does not answer me and takes the burner in his hands and sets to his work.

When we finish, he does not speak and we drift apart once more, tempers cooling. We have done well not to fight. I inspect our work. The calves are sore of themselves, but they will thank us in the long run.

Ballinalee

I have come into Ballinalee to buy some bread and the paper. It is not far from the farm and it provides me with a break from the place. It is a historic village and battles were fought here during various wars. Da was schooled here, and Granny has shopped here all her life. Its name means 'the ford of the mouth of the calves'. It is an odd name and we mostly just call it Bal.

Bridgie from the shop has two freshly made loaves of brown bread, still warm to the touch. I buy them both and take the paper. I will enjoy this at break time.

'How's the farming going?' she asks as she takes my order.

'Not too bad, if only the weather would lift,' I say.

'That will take time,' she says as I hand over the fiver.

'It will surely.'

'Are you writing at all now, John?' she enquires.

'I'm trying,' I lie.

'You'll be our Frank McCourt.'

'He was an old man before he made it!' I laugh.

'No one said it was going to be easy.'

We smile and laugh and I leave the shop, the bread tucked under my arm.

The Purebred

The purebred Charolais cow gave birth a few weeks ago. The calf, a bull, is the culmination of years of selective breeding. He will be a champion – our neighbours have said so too. To him we will give a name, for all purebred bulls must have a title.

He is snow-white and tall, evenly balanced with legs that are straight and not bowed. We do not praise him too much, for fear he might be taken from us by death, or that our luck might turn. Daddy is happy with him. Our neighbour and his friend Rory, who breeds purebreds, come every few days to check the bull's progress.

'Best calf yet, lad,' Rory says each time.

Like so many purebreds, the bull calf's mother has no milk, and so we must find him a surrogate mother. There is the dairy cow we bought for last year's purebred calf, but she has not yet calved herself and so has no milk. Da looked for several days for a foster mother but found none. In the end, a neighbour lent us a cow and she has been providing milk for the new arrival for the last few weeks. She is an old cow and gentle, and though her tits are small he has figured out how best to suck them.

I watch him some mornings as a bubbling white foam emerges from his mouth as he suckles. He is a good grubber. His birth mother never cries for him: she has not the nature of the other cows.

Our neighbour would take nothing for the use of the cow, and so we gave him a hamper with some food and drinks in it and he was thankful. Derek is his name. He is a Protestant and we are Catholic and in the past that would have been a bridge between us, for we have different cultures; he has family in the North, and ours are of this place. But he is an Irishman and a good man, and the walls that were made to divide us no longer exist. The animals do not care what creed we are and they are right in their innocence. Derek has also helped me in the past with difficult lambings, and I have learned a lot from him. I see it as a point of pride now that I do not need to call him any more, for I was a keen student and he a great teacher.

We burned the purebred bull's horns last week and it has put back his progress a pace. It will take time, but he will thrive again. Rory says we should boost his diet by giving him some crunch, which is a type of seed and pellet mixture. Rory understands the genetics of the calf and appreciates him all the more for that.

The purebred bull calf was the product of an artificial insemination, and all we know of the bull that fathered him is that it was British and called the Flying Scotsman. The two shall never meet, but we must pay a royalty for the calf to the sire's owner. We must register the calf with the purebred society and then begin his training with rope and stick and noose. He will be a thoroughbred of cattle. Already Mam has asked that we keep him as our own stock bull.

'And what if we got offered ten grand for him?' Da asks.

'We can cross that bridge if it comes,' she replies.

'That's a big bridge,' he says.

'I invested into that cow, she's partly mine, and I've seen

no return from her yet. The bull calf will stay,' Mam says with an air of finality.

I do not have any share in the cow and so stay quiet. I will let them debate this themselves. It is a happy talk, the talk of money and success. It is like the birth of Hazel all over again, and I know it makes them smile.

When our dairy cow finally calved, we moved quickly and put the pedigree bull calf onto her teat. It took a few days, for he often escaped back to his foster mother, who kept crying for him, and I knew then that their bond had been strong. When the foster cow settled, we took her back to Derek. I wonder now, does she think of him, or look for his scent amongst her herd? There are some things in farming we cannot know.

It's All Greek to Me

In evolution's eyes, the cow has tamed man as much as man has tamed the cow. With the domestication of the aurochs some 10,000 years ago, the Indo-Europeans' diet changed considerably and we turned into milk-drinkers.

Man had previously been lactose-intolerant, as is still evident in nations such as China, where the relationship between man and cow is weak and dairy is not much consumed. With this dietary jump, humanity could now readily absorb vitamin D, thereby giving us stronger, healthier bones and teeth. There are more calories in milk than in crops, which in turn meant a more robust, sturdy constitution. Tending cattle could be a nomadic practice, and so people were no longer tied to one area. In this way, the milk-drinkers could spread to new lands, conquer new territory, and impose their own their culture. Our first great animal relationship was with the cow and not the horse.

It was not a one-sided affair, however, for the cow gained a protector. Man would now guard it from predators and even domesticate other animals to do so. What was the taming of the wolf for, if not to create dogs who could protect our herds?

In Ancient Greece, the geography and mountainous terrain were not suited to the cow, and so the goat and sheep became the dominant domestic animals in the region. Perhaps this is why it was a golden fleece Jason sought and not a golden hide. In nearby Anatolia, there was a cult of the bull, but to the early Greeks perhaps the cow was viewed as a mysterious foreigner. It was in Crete that the attitude towards the cow eventually began to change.

The story goes that in the long ago, Minos, the first king of Crete, sought to justify his sovereignty, and so he prayed to the God Poseidon for a sign that his rule was by divine right. Poseidon sent a snow-white bull out of the sea, on condition that Minos would sacrifice it as an offering of thanks.

But, on seeing the beauty of the animal, Minos could not bring himself to end its life. He sacrificed another bull in its place and gave the white bull a herd of cows. The betrayal hurt Poseidon so, and in anger he bewitched Minos's wife, Pasiphaë, making her fall in love with the beast. The craftsman Daedalus constructed an elaborate fake cow for Pasiphaë to hide inside, so that the bull would be tempted to copulate with her. Pasiphaë bore a monster, half-man, half-bull: the Minotaur. And into it Poseidon poured his anger and cruelty, for the Minotaur ate the flesh of men.

After getting advice from the oracle at Delphi, Minos instructed Daedalus to make a labyrinth in which to keep the Minotaur. When his son Androgeus was killed by the Athenians, Minos waged war in vengeance and when he won, he demanded that every few years seven Athenian youths and seven maidens should be sent into the labyrinth to be devoured by the Minotaur. When the third sacrifice approached, brave Theseus volunteered himself and, unravelling a ball of string from the entrance to the labyrinth, he confronted the Minotaur and killed him with his sword.

Myth and history are strange things, for it seems that sacrifices were actually made in real life: as a vassal of Crete, ancient Athens had to send a certain number of souls for sacrifice. Perhaps they were killed by a bull-headed man, perhaps by priests disguised as such. Whatever or whoever took their lives, the Athenians feared the Minotaur and the cult of the bull.

Thinking of this violent mythical beast so caged, I cannot help but reflect on our own times. Do we not keep our cows locked up in labyrinths of steel now too? I think of the industrial feedlots of the Americas, where hybrid cows are bred, never knowing either grass or sky. They too are removed from instinct, they too have felt the wicked hand of man, they too wait for their day of slaughter. Perhaps the story of the Minotaur, born to a fake cow, is a warning to us now. That we must be careful if we attempt to interfere with nature.

It was, after all, the Minotaur that Dante met on his descent to hell: the first guardian of the walls of Dis, a warning of the violence against nature and the violence of man. Perhaps we must rethink our relationship with that Minotaur, for he is partly us.

Trees

I'm in the fields today. Da has asked me to collect the cut firewood from our land at Ruske's by the crossroads. It is not raining and it is nice to be out for a few hours.

The council recently cut down several trees to make for a clearer view by the road and they have given Dad several bales of sheep wire in exchange. The trees, which were mostly ash and beech, are over a hundred years old. The woodsmen have already cut them into short metre lengths, though their trunks are so thick that it takes strength to lift them and carry them over to the front loader of the tractor. The ground is still wet, so I cannot bring the trailer into the field, for I would track the ground and risk becoming stuck.

So I lift and load and drive back across the road to the house and dump the timber. I have been repeating this for several hours now and my back is tiring. A light mist has fallen but it is not cold and I work in my jumper and hat.

I shout and groan at the bigger cuts of lumber, for they are heavy, but I will not be defeated by them. I like this lifting and amuse myself by imagining that I am some Viking or 'The Mountain' from *Game of Thrones.*

Once, this field had many more trees, but when Ma and Da bought it they cleared much of the ground and it was transformed from a farm that looked Protestant to one that looks Catholic. The old Anglo-Irish farms have beautiful trees and plantations, like in England, while we natives seem to cut the great trees down in favour of small hedges. I do not know why this is, but I have seen it time and time again. It is not something that I much like.

As I work I think of Robin, the old owner of Ruske farm, and how, as I child, I called him Robin Redbreast. How he used to talk to my brother and me over the ditch. How he helped us clear the ragwort from the meadow ground one summer. I think of his fondness for my mother and his wish that she have first refusal on the land when it went for sale. It's strange that fields can have memory, or rather hold memory. I suppose it is like streets in cities, how they can bring back earlier times. On my bookshelf is a short essay from George Orwell in which I have a note to myself from the day it was bought; it reads: 'The day she told you it was over. Bought on Macleay Street, Potts Point, Sydney. Tough day . . .'

I am not sure now if it is the book or the note or the street, but they have all become part of the same memory, like old Robin at the ditch. I realize that I never did read that book, but perhaps that was not its purpose.

My bucket full, I start the tractor once more and drive

back to the yard. I move slowly on the road for I do not want to track dirt or send it flying into the windscreen of cars behind me.

In truth the clearing of Ruske's was a good thing, for Robin had divided and portioned much of the land for sheep and it was no longer right for cows. We needed to remake it to suit our own needs. We had owned the farm next door and I remember the day we broke through the gap and united the property into one holding. We have kept the name Ruske's, though for all I know perhaps strangers call it Connell's now.

I break after the fifth load and take a mug of coffee. My back is covered in sweat and my arms with a fine layer of sawdust. It is a nice dirt. I walk back out to the tractor and dump the load. The mountain of logs is growing and I do not know when we will burn it all, for the house runs on kerosene now, but my sister and brother have solid fuel and should be glad of the wood. It will keep them warm on cold nights.

We let the first batch of lambs out with their mothers last week. They are in the paddocks around the house. It is too early to tell, but they seem to be OK. I hear them call to one another, their cries loud and clear. One must take care when releasing lambs and sheep and do it in small batches, for the mothers may get separated from their lambs and, both becoming confused, they may not recognize one another and a lamb may be abandoned.

I have learned this lesson the hard way, for I released twenty lambs and their mothers a few weeks ago. The ewes ran for the fields, happy to be released, but the lambs had never before left the shed and were slow to venture out. By the time I got them all out, the mothers had run for the ditches and the lambs were left on their own, crying and mourning. I cursed myself then and had to bring the whole lot back in and release them in smaller groups. I have never known a cow to confuse their calf. I suppose the cow is more intelligent, or has a greater sense of smell. But I am still learning with the sheep and must be patient with myself. Da did not know about this mess up. Sometimes it is better to be alone with mistakes.

I cast my eyes around the paddock as I reverse the tractor and head for the road once more. Please God let the weather hold. I hope we do not need to bring the flock in again.

I drive back to the field and load up the trees. I have many more loads to go before it is all home. I will get it all done today, though; I have set myself that chore. I will be tired this evening but it will be a good tiredness. The rain holds off and I am warm in my work. A neighbour waves to me from his tractor and we salute one another, arms raised in greeting like the Celts we are.

Smoking

I've knocked the fags on the head. I used to love the act of lighting up. Especially after the birth of a calf or lamb. I would lean by a gate, inhaling and enjoying the moment, savouring the taste. But no more.

I have the running to blame, I suppose, for it has turned me healthy, but I quit out of shock too. I had come inside from work one day and coughed and coughed, and a small droplet of blood emerged. That has scared me now into total abstinence. I am still young enough to hope that the damage will not have been serious. A sick farmer is no good to anyone. A sick man no good to his family. I know it was right to stop, but I still sometimes miss the way cigarettes gave me an excuse for reflection.

At midday, I sometimes hear the angelus play on the radio. I do not always stop to pray, but the moment of bell ringing is one of meditation. I think of the day so far and simply try and be. I have read a great deal about mindfulness in recent years. It has become a buzzword, I suppose, and yet in it there is truth too. As a farmer, I have to be in the moment, for lives depend on it.

I am lucky in this job that I am able to tune out from the world when I want to. I am not so connected to the world as I once was. I have a phone, yes, but it was bought for me so that I could be contacted when I was away, and I would be happy not to have one. I am getting out of the habit of technology now, and there is freedom in the absence of it. Perhaps Birchview is my Walden. I have felt that it is only in the last year that I have finally begun to live. The thought came to me quite unexpectedly when I was swimming in the local pool. I was in my fortieth or fiftieth lap and I touched the wall, breached for air, and I knew in that moment that I was comfortable in my own skin.

I feared then that it would be a fleeting state and yet, since then, that knowing has only increased, the pelt of peace has only thickened. I feel it best as I run through the forest or cycle down country roads, as I bring a calf or lamb into this world. It is at these times that I feel I am experiencing the sublime, the sacred, the marrow. Perhaps I feel that I only began to live a year ago for then I feared I was to die, but I think it is beyond that. I am no longer content merely to be alive – no, not when there is living to be had.

Sometimes when I go to Dublin to visit friends, our conversation revolves around their phones, they look every few moments at their Messenger or Snapchat or Twitter. I do blame them for this and I once did the very same. It irks me now, but I do not say anything, for I know enough of the rules of cities to know that this would be rude, and so I accept this distraction as their way of life, as they accept mine. I have become, in a way, exotic to them – a farmer, a writer, a man living a different way of life. All I do now is work out and farm, and they might wonder what sort of life this is. To me, it is a lovely one, though I too can see that my terms of reference have changed. I was at a dinner party a few weeks ago with some actor friends and found myself retelling a story about the death of an animal. I realized that it sounded odd to them, though it was normal to me. I am not ashamed of it, the life I have on the farm, as perhaps I would have been years before. It is what feels most real to me now.

Haircuts

Da is clipping the cows' tails today. It's a thankless job, but it saves much hassle later in the season. Outside in the fields, the cow's tail is long and useful, it helps it cool down, keep flies away and acts as a communication tool amongst its fellows. Being in the shed, however, their tail tips have become hardened and matted with shit, and this can provide a source of infection. Indeed once I saw a newborn calf suckle a dirty tail, thinking it a tit; he got sick afterwards, for shit is not meant to be drunk.

Some dairy farmers don't just trim the hair from the end of the tail, they used to actually practice the docking of the tail itself, using an elastic band to cut off circulation to the unwanted part. To dairymen, this painless process meant they had easier and cleaner access to the cows' udders for milking. Thankfully this practice no longer occurs.

Da walks through the shed, knife in fist, and slowly takes the first cow's tail in hand, then cuts the burrs of hair and dung from it bit by bit. Sometimes the cows stand to and let him do his work, while others walk through the shed and he follows after. I think they probably like this beautification, for they do not put up a great fight. You can get a machine now to do this job – a cutting head that attaches to a mechanical drill – and we keep meaning to buy one, but each year we forget.

We move to the weanlings then and it is their haircut time too. We load them in the crush and one by one we run an electric shaver down their backs from head to tail. Next we place a dose of drench on them, which will prevent lice and growths from taking hold. It is bad to see calves infected with scabs and lice. It will not kill them, but, if left long enough, it leads to hair loss from scratching, and, if left further still, the skin dries and hardens and their scratching makes it bleed. It is not a big job but it is a necessary one. Soon the sheds are a litter of hair. The animals have taken on a more manicured look and seem happier, or so we tell ourselves.

'I can give you a buzz cut if you want,' Da says to me at the close.

'Aren't I losing enough as it is?' I say.

We laugh and pack away our tools.

Brother

The purebred bull calf has gained a brother. It was not a planned thing, but such are families made.

It came about after one of our best cows, the Elphin, calved a few days ago. Her udder was huge and overflowing and, try as the calf might, he could not drink all she provided and she contracted mastitis.

The illness is characterized by a swollen, red udder and is caused by milk not being drained from the beast's quarts: it goes sour inside her and creates an infection. If the cow is not milked and the badness removed, her tit will die and she will not bear milk from it again.

In bad cases I have seen the udder rupture and the puss

emerge; in worse cases I have seen the flies take hold and the rot set in. If mastitis strikes, we must act quickly to prevent such suffering.

We treated the cow's tit, milked out the foul-smelling badness, and then injected her with a small dose of antiseptic. It was decided it would be best to get another calf to help keep up with the milk her own calf could not manage, for she was a good, kindly beast and could easily rear twins.

Da set off in his jobbing coat to Carrigallen mart and returned two hours later with a suck calf. The man he had bought him from had said there was nothing wrong with the small bull, save that its mother had been a wild bitch of a heifer and, upon dropping her offspring, had never taken with it, nor let it suck. She too would be sold, he told Da, for no man wants a beast with no nature.

We locked the Elphin in the headlock and prepared ourselves for a long process of many days, for suck calves are often raised on milk replacer and have lost the instinct and neck to suck a cow. And yet the calf went straight for her elder and began to suck her tits almost instantly. He drank for ten minutes, then moved to the next quart.

'Begod,' I said, 'I've never seen a calf as quick. You got a right lad there.'

'He'd want to be, at the price calves were going.'

'How much?'

'Two hundred and fifty.'

'Still, he's not a weakling. He'll make a right bullock.'

The cow kicked him then and we thought the suck calf's luck had changed, but, undeterred, he sucked on.

'He'll do rightly.'

'Thanks be to God.'

The Elphin tolerated her foster child, but would not yet let him suck her without the headlock on and so, three times a day, we put her in the restraint and allowed him to feed. He was brave and strong and so hungry that at times we had to ensure he did not drain her dry and leave no milk for his brother. The work went well until she began to dry up and the milk went from her.

'That new calf is failing,' Mam says as we meet at coffee break.

'How so?'

'I don't think he's getting enough milk any more, have you checked her elder?'

'I'm putting him to her three times a day.'

'The goodness might be going from her milk.'

Mam was right. Her elder had grown smaller and the fountain had turned to a stream; she was but a cow again, a cow fit only to raise one calf.

It was Da who thought to put the suck calf on the dairy cow and thereby give the purebred bull calf a brother. The dairy cow had milk enough to raise two and neither would starve or fail. The transition was an easy one, for the Elphin had never truly bonded with the suck calf and she did not cry or bemoan his loss when we took him to the upper shed. The switch was quick and neither the dairy cow nor the bull calf seemed to mind. The purebred had gained a brother, the dairy cow another calf. We bedded them with fresh straw and left the new family to settle.

'Funny week,' I say to Da as we walk back to the house.

'That's farming.'

We smile and quicken our step. We shall take this small victory.

Writing

My generation was to have been the generation that would stay at home, reaping the benefits of the economic boom they called the Celtic Tiger, but the recession came and we too emigrated to the four corners of the globe. Some of us are returning now, but perhaps we have become unsuited to this place, for we have tasted so much of life. We have returned with new ideas, to find a world that does not want to change. Mam says that to stay here will waste my youth, that I will wake up a bachelor. But at times I think that perhaps I shall never return to a city again, that I will give up the dream of being a writer. I can remake myself as a farmer and settle myself to the land.

When I came home, my plan was to write what I needed to and then get back to the city, but the farm has taken over. I only have the time to make small notes to myself, patches of paragraphs here and there, in between the sheep and cows.

When I was in Sydney, I met the writer who became my mentor. He has written many books and won much acclaim. We are unlikely friends and yet I call him every few weeks and we talk about books. He does not know it, but those talks fire my imagination and put gusto in my belly once more. He once took me to the opera in Sydney. He spoke fluent Italian and understood the words and beauty and explained the story to me so that I could see the beauty in it too. I have not forgotten that day nor all the days where he has slowly guided my hand. I know that I am in my journeyman days now and David understands this, for he too served his time many years ago. He is the master I sought for all those years.

Outside a cow roars loudly and in distress. I put my pen back down and think. I look to my watch and see it is near time to feed the animals again. Words will have to wait.

Failure

The weather has turned once more. For several days now, it has been wet and hailing, and the wind fierce. We have four sheep missing. I counted only forty lambs today. I returned with the dog but he too cannot find them. We are not sure if it was foxes or dogs that took the lambs; perhaps it was the weather, but we can find no bodies.

Da and I are both tired. The nights have taken their toll and I am walking now like a zombie around the yard. I am so tired that I no longer care about the rain, though it pelts my face and stings in its fierceness.

The sheep are in the upper ground, in the rented fields of my uncle Mick. We must bring them towards the small paddocks and then up the lane to the sheep shed.

Genie, my sister Javine, has come to help us move the sheep, for it is an emergency now. We know that if they spend another night outside, death shall come and take more of them. Javine is not a farmer and she does not often come out to the yard. She gives out under her breath, for I know that she does not like the rain and would rather spend her Saturday on her phone or with her friends.

'Where is Daddy? Why isn't he helping?' she asks me.

'He's calling the sheep,' I explain, and we look back to see his small figure shouting out, 'Kitty, kitty, kitty,' which is the call we use for our sheep. Stupid as they are, the sheep have learned that this means food is near and usually come running, but today there is pandemonium. Some have begun the long march to Da, while more stand and wait. The lambs cry out through the sheets of rain. It is getting harder to see now and my glasses are fogging up.

I shout to Da to come up and help us, for his actions are pointless, but he does not hear me and so I must muster them myself. I think perhaps Vinny would be a help, but he is too young still to muster and would only run the sheep. And so I must be both shepherd and dog.

Javine has given up and no longer mushes the sheep on. I must not shout at her, for she is a child, not a farmer, and a row now will cost us the entire operation.

'You're doing great,' I say. She looks up at me and smiles.

'He's still not helping.'

'We'll get there ourselves.'

We march them from the upper ground and down by the river towards the reclaimed paddock. It was once covered in rush and whin until Da and I reseeded it. I know its every acre, for I picked every stone from it by hand myself. The water stands in puddles and splashes as we walk. The river is near bursting and still the rains come.

At last we march the eighty animals into the paddock and quickly shut the gate behind them.

'Genie,' I yell. 'We have to get them onto the lane. Hush them towards the gate.'

She nods and does as I say and we begin the muster once more.

The lambs are confused and cannot find their mothers; the field is noisy with loss. I cannot hear myself think. Slowly, painfully, we move them on, edging towards the gate. We must run back and forth and keep them moving. If only I had a dog.

I bought Da a trained sheepdog a year ago. It cost me a thousand euro and two weeks of hard labour in Canada, but the dog was not worked and grew bored and bold. Da tied him up and that brought on a sort of madness, and in the end the dog ran away. I cannot blame him.

'Come out to the fucking field now and put down that bag,' I yell to Da.

He does not respond and I continue my drive.

'Come out!' I do not know if he can hear me over the rain.

Finally, at the last field, he walks out. Javine is frustrated and has given up and gone back inside. I am alone and march the lambs on. I do not speak, for I am so tired and wet.

'Mind the break,' I shout to Da, gesturing for him to stand by the ditch, lest the sheep should turn. If one sheep breaks, they will all move and perhaps escape and everything will have been for naught.

He holds, as do the sheep, and slowly now we march them to the gap and out across the bridge to the shed. There is much crying and shouts from the ewes as they begin to sort through the young and find their own. The animals are in; it has taken an hour.

I say nothing to Da about not coming out to us sooner. There is no point. The animals are inside – that is all that matters. There would have been a loss tonight, that's all I know. The rain and cold would have claimed another.

'I'll walk the boundary and make sure we have them all,' he says and sets off back into the sleety mists.

I unfurl a bale of hay and fill the plastic container with water. This year has been terrible. The worst I have ever seen. I believe, as do all farmers now, that climate change is real. There has been but one day of frost in the whole winter. The world has changed and we are running to catch up. We, the keepers of the land, can no longer keep it safe. One day the river that runs by the house will overflow and take life with it. We have dredged it to prevent that happening, but we cannot dredge for floods such as these.

The farmers in the west are still under water. They will have to slaughter some of their animals, for there will be not feed enough for them. Athlone, a town south of us, is inundated. The Shannon, the longest river in the British Isles, is out of control and has been reinforced by sandbags. The council is knowingly flooding some areas to prevent the collapse of urban centres. It is a national emergency. We are lucky here, if this is what luck looks like. We shall not drown and for that we must be thankful.

The crop of lambs are delicate and frail, not thriving as they should. Perhaps we lambed too soon. We have enough silage to last for another two months. Things will be better by then. It cannot rain for ever.

I dry my face with a towel and clean my glasses. I cannot see Da, but I know he is out there, searching the ditches. I will go out to him, for it is a horrid task. As I walk out I see him stooped low over the lane. He gained so much from his life of building, but I can suddenly see by his slow gait how much it took in return.

'Anything left out there?'

'Nothing,' he says.

'Fuck them fields. There's no cover out there. We'll have to plant some more trees out there. They need shelter.'

'You'd need a fucking roof over the whole thing.'

'You would.'

I have lost my anger at him. I cannot understand his ways sometimes, but the lambs are safe. I turn that hate towards the weather, something that we can both agree on.

It has been raining for three days straight now. We do not think it possible for much more to fall and yet still it comes.

Granny

Granny is my godmother. Her name is Mary, she's ninety and the head of the Connell family. I call to see her at least once a week. She is funny and intelligent and we talk for hours on each visit: she tells me of her news and of her hens, and we discuss the farm and my writing.

Granny is Da's mother. She says he is very like Granddad, her husband. I never knew Granddad, for he died a few days before I was born. He was called John, which is how I gained my name. Grandda was born in 1890, which means his father was born shortly after the Great Famine of 1845–52. No stories have been passed down from then, perhaps because the pain of the genocide was so pronounced. Some 100,000 people lived in County Longford then; there are only 30,000 now.

I suppose it was no wonder Grandda fought in the Irish War of Independence at the turn of the last century, for the want of freedom, of control, must have been so vivid, so needful then. We had been failed by the old master and it was time to stand up for ourselves.

There are stories of him from those wartime years: rumours of his capture and escape from British custody, the assassination of an informer, his work as an intelligence man. He has always loomed as a hero in my mind. A hero of freedom.

As Granny says, it was a different time, things were done that would not be done now. Granny is the last woman in Ireland to receive the IRA widow's war pension. She has not made this known to people, for this is her business and she is a private woman. She does not speak about the Civil War, which turned brother against brother in the wake of our partial independence from Britain, though I know our family were opposed to the treaty and wanted all of Ireland united. I know too that things occurred then that have caused rifts that still exist in the community. These things are not easily forgotten.

Granny lives with Uncle Davy and his family. They have a busy house, as my uncle runs an undertaking and wedding business. Davy built a funeral parlour a few months ago. He tells me it is the way of the future – that wakes and removals, which have been a part of the fabric of rural Ireland for so many centuries, perhaps since before the time of Christianity, are now a thing of the past. People no longer want to have the body of their loved one in the house for three days. We will do things the American way in the future, he says. I find this odd, for it is our rituals of death that have helped us mourn those who have passed. The three days give us, the living, a chance to grieve. Mam has said that when she dies, she will be laid out in the house and return to earth as a swan on the Camlin River. It is an understanding from the old times, from when we were Celts, before the coming of Christ; we believed our souls, upon dying, moved to the next living thing, be it animal or man. When my time comes, I too shall go in the old ways. To live in the countryside is to accept death as normal; it is not removed or hidden from us, but a part of life, and for that I am thankful.

When I go up to see Granny today, I check on Davy's sheep and see how they are faring. They have had a lamb born in the night and Ellen, my aunt, is out feeding her.

'The weather's still bad,' she says.

'It is that. How's the little one?'

'Good now, nearly sucking on her own.'

'That's great. She's a fine lamb. How are the birds fairing?'

'One has a bad hip. The vet says she might pull through, it's hard to know.'

Davy and his son Jack have collected an assortment of animals over the last year, the most recent being a pair of rheas, which are large, flightless South American birds. Granny calls them the ostriches. Ellen calls them a nuisance.

Between the rheas and the two llamas, Davy's farm is closer to that of Doctor Dolittle. Still, it is a change from the cows and brings a smile to all our faces. Granny talks to all the creatures and they know her now by sight and wait as she brings them each a special treat: lettuce for the rheas, and carrots for the llamas. They know, I think, that she is old, and are never boisterous or gruff when she appears. They know they must behave.

Granny has the kettle boiling when I walk inside and orders me to make a decent mug of tea for myself while she prepares some toasted brown bread and banana.

'Any news?' she asks.

'Nothing, only lies,' I say.

'Then tell me the lies!'

We laugh and sip our tea and talk of the terrible weather. She informs me that the west is still in a bad way. She tells me too that a neighbour has died. I do not know them and tell her as much.

'How would you know them – they've been living in California these last twenty years. But I thought you should know.'

'Will they be a-burying at home?'

'I'd say they will, but I won't go, none of them came to Mick or John's funerals.'

I sense the anger in her voice. No matter their age, both men were still her sons. She had thought she would be dead before them.

'That's fair enough,' I say and we leave that subject.

'How's all going on the farm?'

'Ah, grand. A few lambs this week and a new calf.'

'You're a great help to your father. He was up here the other day and said as much.'

I nod, for he has never admitted that to me.

'He's a busy fool at his age to have so much work on,' she adds.

'He has to do something,' I say.

'I suppose so, but less would do.'

'He's not that sort of man.'

'No, he never was.'

Granny stands up and walks to the press and takes out some biscuits.

'Now, I know you're on the healthy eating, but these are lovely with a bit of jam on them.'

I cannot refuse her and, though I do not eat sweets any more, I take the treat and enjoy it. We shall not always have this time together and I make sure to savour it. Granny has always been there for me.

We talk then of neighbours and gossip. Granny has much news and after an hour or more, when the talk has run out and our criticisms of everyone else's failings are done, we part and agree to do it all again in a few days.

The Black Dog

Da has spotted something in the fields. It looms and roams by the fringes of the sheep that are still out. He has told me to keep an eye: it could be a wild dog come to kill; maybe there is more than one of them. I have not yet seen anything, but I have seen the damage wrought by their packs: whole flocks massacred, limbs torn asunder. They say it takes a rare man to emerge through a dog attack and still farm, for it takes the heart from one so.

Tonight I shall be on the black watch. I will check the fields for signs of it, of them. The shotgun is sitting in the shed, for a week ago Da killed a fox in the upper ground, sure that it had taken one of the missing lambs. He killed him with a single shot, and cut his tail off. There is a den of foxes in the upper ground. If they have a taste for lamb, we are in trouble. I do not like to kill them, but I would have no choice in the face of the slaughter of the young that I have helped bring into the world. The gun sits in the shed, loaded. I must be equally ready when the time comes. I cannot hesitate.

We make sure Vinny is secure in his house at nights now, for he is young and could be led astray by these wild dogs. If he started roaming with them, we would have to put him down, and I could not bear that.

I waken at one to confuse them. For dogs can tell the time and have probably watched and know my usual routine by now. Rory, our neighbour, has told us of a collie he knew in Leitrim, where he is from, who was a killer, but he was clever and washed himself in the river after each kill. It took the farmer weeks to discover the true culprit, by which time ten lambs had gone missing.

The yard is peaceful when I walk out, the cows are resting, the calves asleep. The lambs have huddled together and the ewes we brought back inside are quietly eating. There is no sign of danger, just the sense of it.

I take my lamp and walk the close paddocks. I shine the bright torch and catch the eyes of the few old sheep that are still outside. They are eating in the darkness calmly and baa to be so disturbed.

'We will bring you in too,' I say.

I am not sure if the dogs would enter a shed. Perhaps if they were wild enough, or hungry enough? If they were to come in the shed, then the lambs would have no chance of escape. I would kill whatever animal did that and feel no remorse.

I have only ever shot one thing in my life. It was ten years ago, when we had reseeded a field. The birds kept eating the grain and we feared they would destroy the growth, for they pecked so at the ground. Telling no one, I walked there with the shotgun thrown across my shoulder. I kept downwind of the pigeons and when the time came I flanked them and took aim. I shot and took down a male. I did not take joy in it, but I tied its corpse upon the gate as a warning to the others. The birds did not come back and the seed grew.

This ground, however, has known guns for centuries now. To my left, on Oran Hills, which is the land of our neighbours, the Lees, the British marched the rebels of 1798 to their death in the nearby village. They hung the men from the trees and buried them in quick lime. Da told me that when Pat Reilly bought the ground forty years ago and cut down those same trees, the shackles were still upon the branches; they had become part of nature itself.

Relaxing, untensing, I return to the yard. Vinny is barking wildly and I tell him it is just me, but he will not stop. Then I hear it – the barking of another, a bigger, wilder bark. Da was right.

I find him in the lower shed with the weanlings. He has cornered a bullock and is growling. He did not hear me enter. Turning, he sees me and tries to run, but I am quick and close the shed door and he has no escape now. Faced with me, the animal could turn vicious, and I realize then the gun is not in my hand but in the upper shed and I could well be attacked. I grip the lamp, which is heavy, and prepare to smash his head should he come to attack. It is nature now, pure and raw, and my heart is pumping. The dog bares his teeth and I step forward. I shout and it starts. It does not attack me, nor does it bare its teeth. Instead, it whimpers, cowering, with its tail between its legs. He tries to hide from me and jumps into the weanlings' feeder and I see now for the first time something that is not in Vinny: fear.

I lower the lamp and walk closer. I do not let my guard down, for this all could be an act, a ploy. The dog might make a fool of me and then attack, but he does not. He whimpers and cries and, on moving closer, I see that he has but one eye, and I wonder now what has happened to him. He is no hunter but a broken thing, a stray, who has wandered the night for how long I do not know. He has come in search of food and perhaps has been eating Vinny's feed these last nights.

I take a rope and gently place it over his head and tie him to the gate. I do not know yet what to do with him. The gun is in the shed and I am tempted to kill him. I cannot deny that, for it is easier. We have searched to see the cause of these lamb deaths and now the night has delivered a strange one-eyed dog. I deliberate and pace. I have no one to talk to, for the others are asleep. I must make this decision myself.

The Dream

I have had the most peculiar dream. I do not know if it is tiredness or the lack of sleep, but my dreams have been so vivid of late. I have tried to interpret them as best as I can, and I have looked through dream dictionaries to find their meanings, but for this one there can be no hidden symbolism, for it was so clear.

It began in space, with Da and me as astronauts. I saw the

rays of the sun shining through our shuttle. It reflected upon our suits and masks in its giant yellow glow. I saw the world or worlds down below, I felt the coldness, the vacuum of the outside void and the mechanical whir of our computers and guidance systems. We were travelling somewhere.

After a time, or no time, our lander was touching down. Da was my captain and I his first mate. We worked together and steered the grey craft towards the surface of a world we did not know. It could have been the moon, for it was a lifeless place. We walked upon its surface, smiled and knew our task was complete. And then I do not know what happened, but something went wrong and our oxygen was running low and I knew then that we were in trouble. We raced back to the shuttle.

He sacrificed himself, giving me his oxygen tank so that I might live. I protested, but he would not listen. He took off his helmet, and he looked a younger man, perhaps the age that I am now, his hair black, his face unlined. He smiled then and produced from nowhere a glass of champagne. He raised it to his open lips, smiled to me and, through my protests and laments, he hit the escape button and I shot back to the waiting ship high, high above. I cried and shouted, but he was dead, gone in a moment of joyful agony. He was gone and I was alone.

I woke in fear and sadness. I did not think of our fighting then, only our bond of blood and all the sacrifices he has made in his life for me. I have not told him of the dream, however, for he would think it strange. He thinks me strange enough already.

Lough Gowna

There are places here where the soul of wildness has not been killed. I suppose it is because the land was poor or the water strong, but nature thrives in these places, and in the morning you can wake and think of the dream time, the time when all of this land was forest.

I feel this in the bogs, I feel this in the bottom marsh ground, and I feel this in Lough Gowna. I have come to love this lake, and she has only begun to reveal her secrets to me. By its waters, the swans and seagulls and egrets fish and work. It has islands of oak and ash, including Inch Island, which is a sacred place. Uncle Davy and his son Jack have camped out here and shot and skinned rabbits and caught fish from its shore. A Mulligan man, our neighbour, owns it and ferries his sheep herd across on a small barge. I do not know if he winters them out here; it would be hard work to bring them hay.

There is an old ruin of a monastery upon Inch Island and, as Da says, if it were in Kerry, they would be ferrying tourists, not sheep, to see it. The monastery is over a thousand years old, so old the Vikings attacked it, but we've no tourists here and so it sits idle and unmolested.

The island is full of rabbits, for there are no predators here and they breed easily. Davy told me that a while back a pair of foxes worked out that the rabbits were on the island and were swimming out to kill them. In the end, one of the foxes was caught by a hunter and shot from the shore.

Until recently, the lake people were buried here. This was

their sacred ground, their *reilig*. Their headstones are broken and worn now and hard to read, and wild flowers and nettles grow between them. Their headstones face the rising sun.

I swam out to the island from the shore a few months ago. I did it to see if I could do it, I suppose. I wasn't trying to be Byron crossing the Hellespont; I just did it to feel alive. The water was black but warm and I was not scared until I came to the lake's depths, and then I knew how the crossing foxes felt. Perhaps in some long forgotten past the monks did this crossing too.

Mam's mother's people, the McGaherns, were lake people. She told me that in her childhood the winters were colder and their lake would freeze, and they would move the cattle across the ice to a small island. It is not so cold now with climate change, so I have never seen this happen, but as a child I used to imagine it was as epic a crossing as in the old west in America, and the cows beautiful and calm like a Constable painting.

Cattle Raids

The auroch never lived in Ireland, for not all species made it to this island. It was the Taurine that came here, with the Celts. In the long ago, before the founding of nations, we were a nomadic people, the Maasai of the north, for we traversed these lands through the seasons herding our cattle. It was the cow and not money that was the basis on which all wealth was measured. It was also the cow that gave us our founding myth: the *Táin bó Cuailnge*, or *The Cattle Raid of Cooley*. All my life, I have lived upon the *Táin*. It is our *Beowulf*, the epic of our world, where history and myth converge and where the cow takes centre stage. It was a queen who started this legend, and it was a bull that caused it.

The story begins in a bed chamber, where the Queen of Connacht, Medb, lies with her husband, Ailill. Together they compare their wealth, for in Celtic society the partner with the most would be in control. The value of their possessions is equal but for the fact that Ailill has a white bull, Finnbhennach (a forebear of the real-life White Park cattle breed).

Determined not to cede power to her husband, Medb sends messengers to the Ulster cattle lord Dáire mac Fiachna. Dáire owns a bull, Donn Cuailnge, the Brown Bull of Cooley, the only rival in the land to the white bull.

Being a good man, Dáire agrees to Medb's request and a deal is struck. She may borrow the bull for a year until such time as he breeds for her an heir.

But, as has happened so often in Ireland, drink is the downfall of men, for Medb's messengers boast that, had Dáire not agreed, they would have taken the animal by force. Ireland being Ireland, this news is heard quickly and fully by Dáire and the deal is broken. There will be no bull.

From her seat in modern-day Roscommon, the Queen summons an army to raid Ulster and steal the bull. The *táin bó Cuailnge*, the cattle raid of Cooley, begins.

When I think of the story of Medb, I think of my mother,

for she has built her own small empire through these lands and fields. It was she who suggested to Da that we have a farm in the first place. It was she who invested her profits into fields rather than flats or houses. She has bought cows and bulls and knows their worth. In a man's world, she has proved herself; in a man's world, she has staked her claim. I think too of our neighbour and great friend, Mary Ann Tynan, who runs two farms. She has succeeded where men have failed; she too has built her kingdom.

There have been times at marts when I have looked at the faces of the farmers there and seen the traces of our mythic past. The surroundings have changed, but the people have stayed the same. Our relationship with the beasts has stayed the same, too: they are still the foundation of our rural world, our link with nature, and our livelihood.

Medb's raiding party moved through the midlands, through our home of Longford, through the very fields of Birchview, and from there to Louth and the lair of the bull Donn Cuailnge.

The men of Ulster could not defend themselves from Medb's forces, for they had been cursed by a goddess in vengeance for the king's cruelty towards her while she was heavily pregnant, and were stricken down with the pains of childbirth at their greatest hour of need. Despite her advantages, Medb's raid was a failure, for a spirit called the Morrígan appeared to the bull in the form of a crow, telling it to flee and evade capture. In his rampage, the brown bull of Cuailnge killed many Connacht men.

Ulster lay prone and bare and there was but one man

who could defend them: Cúchulainn, the Hound of Ulster, who gained his title as a child when he killed a guard-dog in self-defence and then offered to serve in its place until a replacement could be reared. Now, still only seventeen years old, Cúchulainn found himself standing alone against the force. He invoked the ancient right to single combat and faced Medb's army one by one, day after day.

During this time, Cúchulainn successfully faced the Morrígan too, who appeared to him in various forms, some alluring, some violent, including as a stampeding cow. It is, however, his fight with his best friend and foster brother, Ferdiad, that is remembered by every Irish school child. Ferdiad was the champion of Connacht, the bravest and best of men of that province, and through Medb's guile and cunning he was forced to fight his beloved brother. For three days and nights the pair faced one another, with swords, spears, lances and shields. In the end Cúchulainn slayed his friend, though it was with great sadness.

It is said he shouted as the final mortal blow to his brother was delivered, 'Thou to die, I to remain. Ever sad our long farewell.'

In their tragic battle, I see a part of our own history reflected, from the Civil War up unto the modern day and the bloodshed of Northern Ireland, with its war and bombs and terrible violence on both sides.

After several months of man-to-man fighting, Cúchulainn pitted against Medb's warriors, the men of Ulster finally began to rise from their pain and slumber, the army mustered and with that the final battle began. Medb's forces lost the day but they did manage to take with them the brown bull.

Finnbhennach the White, and Donn Cuailnge the Brown met at last and began an epic battle themselves. In a long and brutal fight, Donn Cuailnge killed his foe, but in the doing was mortally wounded himself. He then wandered around Ireland before finally returning home and succumbing to his wounds. And so ends the great Ulster Cycle.

The story is a part of my life, a part of every cattle man's life in this country. Revisiting it now, I see perhaps it is not just some epic tale of war and bloodshed, but a morality lesson on the covetous nature of man and the worthlessness of possessions. In it we also see that animals have their own intentions, for Donn Cuailnge, despite whatever the Morrígan did, did not wish to be dominated and fought against his imprisonment.

Interestingly, the *Táin* was written down for the first time in the twelfth century in the *Book of the Dun Cow*, the *Lebor na hUidre*, so named because the manuscript was said to be made from a dun (meaning greyish-brown) cow hide. In it are contained the legends of pagan Ireland as well as Christian works of faith and history pieces. The *Book of the Dun Cow* is the earliest extant Gaelic manuscript and without it we should not have so beautiful a telling of the *Táin*. It shows what great respect these Christian monks held for the old stories. Even as men of a new god, they did not cast them aside as mere idolatrous tales.

How much of this myth is based in real events, we can only guess. How true were the deeds of Perseus or Heracles? And yet there is always some fact in fiction.

It is strange to talk and think of cattle raids in the twenty-first century, but in recent weeks I have heard news of several cattle raids in the area. It is the worst crime that we farmers can imagine. Since the collapse of the Celtic Tiger some five years ago, lawlessness has increased in rural Ireland. People no longer feel safe in their homes, for gangs of thieves roam the once-quiet countryside. I do not know a farmer without a gun now, for the police are fewer and we are all but alone. In the last three years, over 10,000 cattle have been stolen. The raiders are most active along the border with Northern Ireland, where the warriors raged in the long ago. In our hardship, we have resorted to the old ways. The stolen cattle are butchered hastily and their meat sold; few are ever recovered alive. Their bones are often found dumped on lonely back roads. The *Táin* is not dead, it has but changed its form.

Bloat

Today I am glad of my phone. In the shed this morning a lamb that has always been somewhat poorly is lying out. He is foaming from his mouth and panting. His gut is swollen and large and I do not know what is wrong. This is a new illness to me. I am alone out here and the lamb is in pain. Dad will not know what to do, for we have only had sheep these past three years and it is to me he looks for insight on these beasts.

The lamb's mother paws the ground in an effort to make him stand, but he will not rise. I set him upright, but after a few steps he collapses once more. I do not know what to do. When I squeeze his gut, there seems to be air, a huge quantity of it stuck inside him. It is some form of bloat, that much I piece together. To be a farmer is to be a student for ever, for each day brings something new.

I Google the symptoms and find the cause and cure. I sigh gently, for it seems the illness is treatable here without the need of a vet. This lamb has had a hard road already, for his mother had only one tit and he has been fostered to a surrogate. She and he had taken well together and formed a strong bond, but now it seems fate has dealt him another bad card.

I walk inside to get some vegetable oil and baking soda to quell the bacteria in his gut, as the Internet has told me.

'Got a lamb with bloat,' I say.

Mam and Da are sitting in the kitchen. They've been having a serious chat and the news lifts them out of themselves.

'Swollen as big as a balloon,' I say.

'Which one?' Mam asks.

'The fostered lad in the top pen.'

'He was always a bit ropey,' Da says.

'Leave it with me.'

I must stomach-tube the mixture to him. The article has said that natural yoghurt will also work. I grab some from the fridge and return outside with my home remedy.

He is panting heavier now and I worry is it too late and that he may die. The article has said that's a possibility. I do not know how long he has been this way. Perhaps I missed him in my rounds. Perhaps Da did too.

When running our night and day shifts, we leave notes by the front door for one another. Looking back through those scraps of paper recalls each of the late nights and long evenings, the sicknesses and the success. Of this lamb, there is nothing. A blank record.

I insert the slender plastic stomach tube down his throat and carefully attach the syringe full of fluid. If I get this wrong I will send the fluid to his lungs and kill him. He squirms and cries and I hold him tight. I push and inject the fluid.

The article tells me that after a few minutes I should insert a rubber tube down his throat once more to allow the gas in his stomach to escape. I must massage his belly and gently force the air from him. I let the lamb recover and take a moment to settle myself. I do not think he will die, but I must be careful. They say farming has changed, that it has become industrial and mechanized, but still if the farmer has not the nature to care for his beasts when sick, they will die. I have known the hardest of men to be soft and gentle with their animals in a way they never are with their own families.

Settled now, I insert the tube once more, and rub and massage his distended belly. I hear the rift of gas and smile, for the operation is working. He is still swollen but perhaps less so now. I massage and rub and shake him for ten minutes or more. His wool is still short and bobbly, his body is not muscular like the other lambs, but he is by no means a cripple. He has grown with his foster mother. She cries out for him now.

'In a minute. I'm not taking him away for ever,' I tell her, but she does not listen and still cries out.

I think I am finished now, for I can hear no more air. He is still swollen but perhaps the mixture and massaging has worked. I check through the article again, reading to the end. I must isolate him from his mother, for her milk will only feed the bacteria growing in his stomach, and that needs to be killed. He must fast and then I will reintroduce healthy bacteria through the natural yoghurt.

It will be twenty-four hours before I know if I have succeeded. The article tells me that in extreme cases a needle should be inserted through the abdomen into the stomach for an emergency release of the gas. I am not confident enough to do this yet and it would cost 50 euro to take him to the vet, more than the lamb is worth. I must take my chances. I must gamble and wait. If he has not improved in a few hours, I will intervene and perform the emergency operation.

I place him in our 'hot box', which is a small wooden chest below the work kitchen. It has a red heat lamp to keep him warm and safe. I close the latch. I have done all I can in this hour of need and there are other jobs to do.

I have had good luck with the lambs this year so far. I have lost few and those that have died have died of causes that man could not prevent. There was the lamb taken from me by his mother's carelessness, for she lay on top of him and smothered him in the night. Then there was the lamb who was born a triplet and a cripple. Da said it was from the way he had lain in the womb, for his neck was for ever bent to one side and he could not stand upright. I dubbed him Richard III, after that crippled king, and together we waged against death and for several days we prevailed. Through careful massage, his neck began to unstiffen and with aid he could stand. I had hopes for him that he might strengthen and become instead a Hotspur, but hope is a chancy thing with animals; it flew like swallows' wings and left the morning I found him dead upon the floor of the hot box. He and I had lived together for five days. It is dangerous to name animals, for in a name we build a bond.

Those two have been all my losses and I cannot complain; I have heard of far more death in a season and I have seen far more in other years. If I can mend the bloated lamb, I will be happy.

I have lost no calves and for that I am the most thankful. Mam jokes that we have Saint Francis working on overtime at the moment. Let not the saints get tired, nor the older gods. We will pray to whatever keeps luck alive.

At 2 a.m., I waken and do my rounds. I check the lamb in the hot box and find him standing upright. He is brighter now and cries out to me for food. His stomach is still swollen, but he seems not to be in pain and I am thankful for the Internet: today it has served a use and helped me greatly. I want to tell Da of my success but he is asleep now. He will see in the morning. The lamb roars to me and I agree that, yes, you can feed. I feed him some natural yoghurt and then make a small bottle of milk replacer and feed him, childlike, in my arms.

'You're a good lad,' I say and rub his head.

I close the box and walk back into the farmhouse happier now. The day has ended well. We shall see what the morning brings. I boil the kettle and have a mug of tea and a scone before returning to sleep.

The Stock Exchange

He has died in the night. I rise at 9.30 a.m. Mam had forgot to look into the hot box on her morning rounds, so it is I who find him. The foam had come back, it was around his mouth and still wet. His body was stiff and cold.

I know now that I should have performed the emergency needle operation. I had been too much of a coward to do it and now he is dead. It is my fault.

The death has brought back a memory of previous cowardice. I was eight or nine and cycling into the local village, by the bridge over the Camlin River, when I saw a small animal cowering by the roadside. I stopped and discovered an otter. What he was attempting to do, I do not know – perhaps cross to the other side. The animal looked scared and young and something in me said it needed help. I was afraid then to pick it up in case it would bite me. I pushed it off the road with my foot for fear, nudging it out of the way of traffic. Eventually, unsure of what to do and scared to handle it, I cycled home to get a cardboard box to load the animal and bring it to safety.

I did not cry when I returned and saw its lifeless form on the roadside. My young mind understood that a car had killed it, crushing its head. My cowardice had cost a life that day. I have never forgotten that beautiful creature.

Da is not sympathetic when I tell him about the lamb and says it was never fuckin' right in the first place. I tell him of the needle and the ifs and buts, but it is pointless now.

I take a needle and perform the operation on his lifeless body so I will know how to do this for the next time. I pierce the flesh in his abdomen where I imagine his stomach is. I feel the metal push through the layers and then the gas escapes and his body deflates. It was this easy. I am a fool.

I place his stiff body in the empty fertilizer bag and I will take him to the knackery yard later, which is where the dead must go. I am annoyed and disappointed. I am reminded of a neighbour's saying: where there's livestock, there's deadstock. The phrase helps but does not bring back the lamb.

Cash Cows

The lower shed is all but empty. Most of the weanling calves have been taken to market by Da and my brother. They had reached the fourteenth month of their lives, had put on the required weight, and Da decided it was time to sell them. There has been a row and we are not talking. Our tongues, which were venomous a few days ago, have been stilled with anger.

I watched the procession leave the yard all morning, the trailer full of the roaring of the bullocks, the wails of the heifers; they would not be coming home again. Derek, our neighbour, helped with the last load of weanlings, for the day was well into itself and Da feared he would not get good sale numbers. The numbers are a gamble: if you arrive too early, you will not have enough buyers at the sale; too late and good prices have already been and gone.

There are just four weanlings left now, too young to sell; we will fatten them on spring grass soon. Walking through the shed, I miss the presence of the others, their noise and smell and rumble. But they are the payment to the bank for the land. They are money embodied, nothing more. That is what Da says.

On this we do not agree. I cannot see them just as products. They are animals, not mere steak-holders. They may carry flesh, but they carry personality too, memories, and feelings. But to go down this route is not businesslike. And farming above all is a business, I am told.

The reality of beef farming is that the cows live so that they can be killed. They are here so that they may die. If we did not eat meat, they would not exist, or not in such great numbers. All of our cows on this farm will be killed some time or other; they shall get old, or reach their weight, and all shall know the butcher's knife. But even knowing this, and even for the businessman-farmer, I do not believe it is solely about the money, nor that he sees the animals only as future beef. If it were, I do not think he should get up so instinctively in the middle of the night to deliver a new calf, or tend to a sick lamb. There must be nature in the man for the beast, nurturing in the human for the non-human.

In the cities, man has divorced himself from nature. Perhaps it was not meant that way at first, but the separation has occurred almost wholly now, and the most urban dwellers now see is a park, a manicured replica of nature – alive, yes, but carefully cultivated and controlled. There are animals in the city, but apart from the birds and the vermin, they do not roam free. The city dwellers pay us to maintain that link to nature, and we, the growers, harvest what they cannot. Sadly the majority of us now upon this blue planet have lost our relationship to nature.

It was Toronto that brought this home to me most sharply. For nearly two years I lived in its concrete embrace. The trips I took with my then partner to the countryside, to 'cottage country', as Canadians call it, were like an oasis – to be reunited with nature and trees and calm and birds was a sustenance that I needed. It was here I saw bald eagles and bears, salmon and trout in the river, and moose and deer in the woods.

It was not all bad in the city, of course: there was theatre and discos, gyms and cafés, restaurants and young people. But whilst living my condo life, part of me missed these cows and this way of being. It was a sort of *uaigneas*, or loneliness, that I could not fully articulate. It is now that I think that life was meant to be shared with animals, not just other people.

And so these beasts, these cows, are not mere products to me; they are my fellows. It was we who brought them from the wilderness to join our family and walk by our side, and I am glad to walk with them still.

Da comes back from the mart. The prices we got for the weanling calves were not so good as last year and for that he is a bit annoyed. It was not the quality of the cattle, for they were good beasts. He hands the sale docket to Mam and she inspects the prices, calling out the numbers, to which Dad responds which calf it was.

'And for the good red lad?'

'Seven hundred,' Da responds.

'I thought you'd get more for him.'

I can see the hurt on Da's face, though he tries not to let it show.

'Prices are back.'

'Is everyone selling now?' I ask. It is the first words we have spoken in a few days.

'Ah, the whole country decided to come out at the same time.'

'They're good prices, considering,' I say.

'Ah, they're all right,' he replies.

The factories have a monopoly on cattle prices, for there are only a handful of slaughter companies. Larry Goodman is the biggest, with plants in Ireland and the UK. They call him a beef baron. The factories are not the friends of farmers. They know that at this time each year the weanlings are nearly ready for market, feed is running low and bills must be paid. And so, some say, the factories control prices and keep them low, and farmers do not get a fair amount. It is not fair, but it is business.

I have only been to the slaughterhouses a few times. Each time I have found them cold and clinical, for death is their business. Da always says that there is a moment, a second, when the beast is no more and the carcass has replaced it. In photography, it is called the decisive moment: the second before the footstep hits that puddle, or Capa's dying Spaniard hits the ground as a corpse. It is the same in the slaughterhouses: life becomes flesh. Of where the spirit goes, I do not know.

'Did you eat there?' I ask.

'No.'

'I'll put on something.'

I fry a rasher and mushrooms and serve them to him in a sandwich with a mug of tea. I leave the tea bag in, for it is the way he likes it best. It is a small act of reparation. We forgive each other silently for the row and nothing more is said. It has been a long day for Da and he is tired. Mam smiles as I serve him his meal. This will be their twenty-fourth season of selling the weanlings. The ritual has not changed.

Westerns

Like most men of his generation, my father is a great lover of westerns. He has seen them all. Here in the midlands, we play country-and-western music. Men have built careers on it. This Irish country and western, as it is called, has its own sound. It is not so harsh or so blues driven, and it has a soul of its own. It is hard to imagine the barren landscapes of the rugged American plains in this green place, but we picture ourselves as stars in our own westerns. The cows become steers, the horses steeds, and the farmers ranchers.

In these lawless days, we are perhaps closer to the Wild West: we too have frontiers, there are cowboys here, and rogues and villains. We know a local man who shot two thieves that broke into his property. He is not ashamed of it, for they were armed too. The police are mostly absent, so men have taken things into their own hands, and each family is now an outpost. These times have made the country people harder.

Our family has not been spared either: we have been robbed, and a house we owned burned down, and the police were of no use. There are roaming gangs on the country roads and neighbours keep watch and text one another of suspicious activity. Uncle Davy's brother-in-law had his four-wheel quad stolen from in front of his house two weeks ago. It had cost him ten grand and was not yet insured. The thieves tried to blackmail him and offered to sell it back to him for four grand. He is going to lay a trap for them. I am not sure if it will work.

Davy knows the King of the Travellers, who are our native gypsies, for they deal with him for their burials. We are somewhat safe from gypsies, for they will not go against Davy and by extension us, for that could cause a taboo on their families. They will not break their rules of death, for they are very devout. But the king has no say on the roaming gangs. He is powerless and can only police his own people.

Perhaps it is this climate of chaos, but I have been thinking of writing a western and have bought some old classic films. Da and I have watched them together and we discuss them out in the yard between chores.

'That Charles Bronson is some man,' I say.

'A classic.'

'Had you seen that one before, Da?'

'Years ago. I had it forgot.'

'The bit with the harmonica was amazing.'

We are happy in our talk.

Every man has hidden depths and everyone at some point can surprise you completely. It happened with Da some years ago now. I had been learning guitar for some time and had just got into Bob Dylan. Following in the footsteps of so many teenagers, I bought a harmonica and began to practise badly. One Saturday, Da and I were sitting inside after our chores had been done. It was then that he asked me for the mouth organ. He put it to his lips and blew and no sound came, and I made to show him how to use it.

He nodded his head and then ran the blades across his lips once more and began to play such music as the house had never heard before: blues first, old and timeless. I stood smiling, lost for words. He played country and western then, and again his notes were sonorous and clear. In the melody, I could see the great plains, the bars, the brawls, the buffalo and the blood. I looked around me, willing there to be another present to see this, but there was no one but me, an audience of one. He played for five or ten minutes more, then put the instrument down gently.

'It's been a while since I did that,' he said simply.

'That was amazing.'

I have never seen him play again. My father the cowboy.

Do Cows Dream of Electric Sheep?

The cows spend a great deal of the day sitting down in the shed. This is not out of laziness but of nature, for to digest their food they must sit and regurgitate their fodder and chew it once more to fully break down the food. This is known as chewing the cud, and a cow can put in many hours at this activity. I was told once that it also gives the animal something to do and so prevents boredom. Cows, unlike pigs, can amuse themselves.

As a result of so much sitting in their cubicles, their hair has become matted with dried dung and muck and they have taken on a lizard-like appearance, with scales for skin instead of a pelt. It is a peculiar thing to see, but I do not try and remove the hard hair, for it will go once they are outside again.

Cows only sleep for about a few hours a day, and even then not all in one go, but in snatches. It is, I suppose, an adaptive trait to help them stay semi-alert and ready to flee should predators appear. At night, the cow sleeps as the calf does, wrapped onto itself and nuzzling its head upon its flanks. It is said that cows enter the REM sleep state just as humans do; this is where dreams are born.

Walking through the shed, I see their closed eyes flutter, I see their barrel chests rise and fall and occasionally a jitter of a leg or shiver of their flank. Of what they dream, I do not know. Of being outside in the fields perhaps? I know that Vinny dreams of running, for I have seen his legs move that way in his sleep, but a dog is easier to understand than a cow. I walk slowly through the shed, for I do not want to rouse them, for a sleep-deprived cow is a wicked cow and might lash out at her fellows or at us. I leave a small light on so I can see that everything is OK.

Occasionally they waken, no matter how quiet I am, and slowly stand to their feet.

'It's OK,' I say softly. 'Go on back.'

They are slow and cumbersome getting up, but the fact that they can stand makes me happy. I have seen cows 'go down' more than once. It happens during labour when the calf is too big and the mother suffers small ruptures or damage to her pelvis or nerves. If a cow goes down in labour, we are quick to make her stand, for she might not get up again. There are times they recover and the shock wears off and they are fine, but on other occasions shock gives way to inactivity. If they do not stand after three days, the muscles of their legs begin to waste and then they shall never be able to stand.

Years ago now, we had this happen. She was a black cow. I was but a boy then and did not fully understand her true ailments. In the end, we lifted her with hoists and ropes and she stood for a time, and we congratulated ourselves that she would be OK. But when she tried to walk, we knew it was too late – her leg muscles had faded away. She limped to the drinker and sat down, and never stood again. She was culled from the herd two days later. We called her the Little Black.

I remembered this cow years later when I was twenty-one and a homesick exchange student in Sydney. I had taken a writing class as part of my journalism studies and my teacher in Australia urged me on to write something true and of myself. I wrote the story of the Little Black and, in thinking then of my childhood and the farm, my homesickness carried across the page. It was a cow that gave me my first taste of being a writer.

As I walk through the black, dark night, I listen to Philip Glass on my headphones. Up above I can see the stars. It is one of the most wondrous things of living here in the countryside, for on these night shifts I have seen the heavens pour out above me. I have seen the stars and the dust around them. I have seen the hunting moon of January, Venus shining brightly, and the Milky Way as I have never seen it before.

The ancient Egyptians believed that the Milky Way was a pool of cow's milk made by Bat, the cow goddess of fertility. On the nights when I have been tending to the cows and I look up, I cannot help but think I am part of some ancient ritual, that the act of milking a newly calved mother is in some way connected to the birth of whole galaxies in some far reach of this universe. We live our lives – the cows and I, and Ma and Da, too – under this Milky Way. Perhaps it is of those celestial bodies that the animals sometimes dream.

Holding

The cows are faring well at the moment, but Da has said he is unhappy with this year's crop of calves: there are few stand-out calves. The stock bull is not breeding so well. Mam and he have talked and think that perhaps it is time for a change. From now on, Da has decided to use outside sperm to artificially inseminate the cows that come into heat instead of letting the bull to them.

Uncle Paul, the former mayor of the county, was once an artificial insemination, or AI, man and he has agreed to do this for us. Since retiring from politics, Paul has more free time and we are happy to see him. A year ago, Da and Rory, our neighbour, went halves on a nitrogen flask in which to store the straws of sperm and keep them fresh.

A cow will come into heat every twenty-one days in what is known as oestrous. At this time she will exhibit clear signs, she will stand and present herself. Her sisters will rise on her and attempt to cover her, for I suppose the pheromones grow so strong they feel they must act. She will be 'a-bulling', as we call it, true and proper, for only twenty-four hours, during which time she is most receptive and the sperm will most likely take with her, so it is in this time that we must act.

In preparation for insemination, we bring her to the lower shed and place her in the headlock. A cow in heat can be a bit more temperamental than normal, for she is full of hormones.

At two, Paul arrives, and we have a flask of warm water waiting. The water must be at the right temperature to defrost the frozen sperm: too hot and the sperm will die, too cold and they will not thaw out.

Paul removes the straw of sperm from the liquid nitrogen flask and quickly places it in the water. After some forty seconds, the straw is then placed in an AI gun. Paul slips on a plastic glove and inserts his hand up the cow's rectum, from where he will direct the straw as it is inserted into the beast's vulva. When everything is in place, he presses the gun and the sperm is released. It is a precise and delicate process, for if it is not in the uterus the sperm will die and no fertilization shall take place. Every straw costs money and we cannot waste them.

He nods, then smiles to us, for he has done the job. He releases himself from the cow's embrace and she moos gently. The operation is over.

'She should hold,' he says.

'What bull did you put on?' I ask.

'CF52,' Da says.

'He was a great one,' says Paul.

He cleans himself, removing the sodden plastic glove and disposing of the used needle. We will leave the cow on her own for a few hours and mark her time on the calendar. We shall not know until next month if she has taken. If she does not come into heat then, we know it has been a success. In this, as in so many things on the farm, we must be patient.

We do not use AI so much on our farm, but in American feedlots it is now standard for entire herds of hundreds of beef cattle to be put in calf by this method, often using semen from a single bull. Some breeding bulls are so successful that they can alter the genetic fabric of a breed. Sadly for the bull, the semen is collected by hand and he remains a virgin all his life.

Our job done, we relax and hope that the sperm will do their work. Paul and Da share a small whiskey and talk of politics.

The Store

I have run out of nuts for the ewes and must go to the store. We have been buying from a place in the nearby town, but today Da says it would be better to go to McKeon. He is a local man who has set up a store in the parish and we admire his entrepreneurial streak. His business has taken off and he now stocks feed for all animals, as well as fertilizer and feeding troughs and barriers.

'Ten should do,' I say.

Da agrees and I set off in the jeep through the village and up Soran Hill. The Connells came to Longford with the Irish Rebellion of 1798. We had been rebels under Theobald Wolfe Tone, the leader of the United Irishmen. The rebellion had been one born of the want for freedom inspired by both the French and American revolutions, but which ultimately turned into a bloody and brutal affair. Old Connell, so the story goes, travelled up from the south of the country with the rebels, leading him to the final decisive battle in Ballinamuck in the north of the county. The British forces resoundingly defeated the rebels and with it that generation's dreams of freedom.

Granny still talks of that time. She surprised me one day a few weeks ago with the story of Hempenstall. He was the British lieutenant in charge of suppressing the United Irishmen in this area during the 1798 rebellion. He was so big that he could hang a man over his shoulder until dead and became known as 'the walking gallows'.

Hempenstall was so effective at his work that he commanded a platoon of men known as the 'Terrors of the Midwest'. When the Battle of Ballinamuck was lost, the local rebel leaders lead a small force of men to the town of Granard. There they faced the Terrors and, in a twist that only history and not writers can allow, this giant of the British was met by the biggest Irishman in the county, O'Farrell, who himself stood at seven feet high. O'Farrell beat Hempenstall in a fist fight, but the rebels lost the day and with that the war.

The defeated men were then tortured and many were trampled to death by a pack of bullocks driven by the British. Those who survived the trampling were roped by the neck and slung over Hempenstall's shoulder. These things were not recorded in the letters of General Cornwallis, Lord Lieutenant of Ireland and Commander-in-Chief of the British forces, but they have been passed down through the generations. These inherited memories are in the mind of a ninety-year-old woman still.

The jeep chugs up the hill and my Bob Marley CD skips a beat as I brake quickly and take the turn for the feed store. McKeon himself is not there and one of his men takes my order and helps me load the ten bags of nuts. They will keep the sheep going for another few weeks, for we are still only halfway through the lambing.

Next year I hope to perform a controlled pregnancy, called sponging, on the sheep. It brings the ewes into heat at the same time, so we can then put them with the rams and, five months later, all the births will take place during the same week. It makes for a busy but shorter lambing period. This drawn-out lambing is too hard on both man and beast. And while we are coping well now, tiredness is lurking and our luck may turn.

The nuts are added to our account and I take the docket and leave. I drive home, unload the bags and prepare for the evening feeding.

Emergency

I have been out running in the forest. I had only planned to come for a short five-kilometre run, but I have been running for two or more hours – the distance of a half marathon – and my legs are sore. The day is bright and I am making the most of the sunshine.

The sun is shining lower and I must get ready for the evening feeding. I face for home, taking my final turn, and head back through the forest to the car park.

Out of breath now, I walk slowly towards the Jeep. I am thirsty and tired and looking forward to a good hot shower.

I check my phone and there are many, many missed calls from Da and I know in that moment something is wrong. He only calls me when there is an emergency or he is looking for something. I curse myself for not taking the phone on my run, for I see now the calls were made through the last two hours. Perhaps something has happened to him? To Mam? I have a sudden image of Mam finding Da stretched out in the yard, dead from a heart attack.

I make to call him but my phone dies and now I have the added panic of no connection. I turn on the phone once more in the hope there is a little battery left. I call and hear it ring.

'Come home now, you fucking bollox. Where the fuck are you?'

'What? What's wrong?' I have not the time to thank God that he has not had a heart attack.

'I've a calf about to die. Get back here!'

He hangs up then. My heart is racing. There will be a row. I had told him I was going for a run, but he is in a rage now and there will be no use in talking to him of the past. My heart is still racing as I drive back quickly through the country roads. It will be fifteen minutes before I am home and I know that he will stew and stew in that time. His words have undone the calmness of my run, and I can feel the stress within my body.

I return home and find him so enraged that he will not speak with me properly. I can tell he is now a prisoner to his anger.

'What's wrong?' I ask once more.

'There's only a fucking calf dying out here and you never even noticed him.'

'Which one?'

He does not answer me and grunts.

'What do you want me to do?'

'Put the fucking trailer on the Jeep and take the calf to the vet before he is dead.'

'Calm down,' I say.

'Just put the trailer on!'

'Calm fucking down, I still don't know what calf you are talking about.'

He will not answer me.

I drive the Jeep up the yard and attach the small blue sheep trailer. It is big enough to take a calf. I drive back to the shed, reverse the jeep and trailer in the door and walk out. I will not lose my cool. I must be the eye of the storm. It is better not to speak now, for every word will be used and thrown against me.

Da is waiting in the calf creep. It is the big red calf that had the scour all those weeks ago. He is lying curled up. He had been this way this morning, but I had thought him sleeping. Together now we push the calf up and I load him into the trailer myself. He does not help me and I know blames me for this illness.

I do not ask which vet to go to. I shall go to Gormley, for he is the curer of calves. He is old and wise and has saved many beasts for us. It is only on my drive away that the thought occurs to me, why did Da not call Davy and borrow his jeep if the calf was that sick? Or why did he not arrange to get the vet to come if it was so bad? Perhaps he is angry at himself for missing the calf, perhaps he has only just come and found the calf? I do not know. All that matters now is that I am going to the vet. All that matters now is that he lives.

Gormley has been our family vet for many years. He was old in my childhood and is near eighty now. He is of the old breed of vets. Almost a figure out of time he is, sharp, with a vast intellect, and educated better than most of the men he meets. He is our James Herriot and Siegfried Farnon rolled into one. The initial panic now over, I settle myself.

'It cannot be all so bad, Red?' I say aloud to the sick calf behind me. 'It cannot be.'

Gormley is not at his practice when I arrive. His daughter tells me he will be there soon. I will wait now. I try to make to call Da to tell him I have arrived but my phone is fully dead now. It will not turn on.

The practice consists of Gormley alone. It is odd, but he has never taken on an apprentice, nor expanded. When he is gone, there will be no other to take his place. Perhaps that is why he has worked so long. To stop now would be the end of him and all he has built. I have known Gormley since I was a boy, but we have never truly talked, he and I, for he is a man of few words and, to him, I am but another farmer's son, another call in the night. Or I am not even that: I am the beast's illness and how it must be treated. I should like to talk with him of other things, but I know we never shall. That is the way it is and must be.

He arrives at half four and I tell him my problems. He nods and asks to inspect the calf. He takes his stethoscope from around his neck and waits while I open the trailer door and wrestle the calf out. The calf does not put up much of a fight, though there is still vigour in him.

'How long has he been like this?'

'We only noticed today.'

'How old is he?'

'Three months.'

'Is he sucking his mother?'

'I saw him suck yesterday.'

He moves his hands along the beast's legs, sighs and clucks his tongue.

'He has an infection, it's got a bad hold of him, he's had it for a while. Has he been sick?'

'He had scour a while back but I treated him and it cleared.'

'He probably picked it up then.'

'Will he be OK? Will he live?'

'He should be fine once I've treated him.'

I sigh and smile. It is not so bad, Red, we will get you fixed up. The stress begins to leave my body. Gormley tells me to hold the calf while he prepares an injection. He walks into his dispensary. I never ask what medicines he gives, for they are often of his own making. I have seen him mix old bottles with powders and solutions to create some magic life-restoring brews.

He returns and injects the calf and pats him on the head.

'That's a long-acting solution; he should be fine in a few days. If he hasn't started to improve in three days, come and see me again.'

'Thank you, Brian.'

'That will do,' he replies and leaves me. I load Red back into the trailer.

He will add the visit to our account. It will be over 60 euro or more. The calf is a good calf so it will be worth the effort. When I return home, Da is not in the yard. I unload the calf and return him to the creep. Rest is all that can be given now. His mother cries out to him and he responds. He seems brighter, I think. I cannot be sure, but I think he is not going to die. The vet has said as much. He will pull through.

I will not mention the row to Ma. There was a time I told her of every fight, but there is no point any more. Da has not foddered the animals and so I must now begin the feeding. I do not know where he is. It is better I do not know.

Darkness into Light

Red is alive. It has been three days and I am hopeful. I watch him now every time I come out here. I have taken the blame upon myself and think perhaps Da was right to be angry. I had let my focus slip with the calves. I was so busy with the sheep that I walked by them and had not really looked, but I must watch for the signs. Another calf has come down with scour, so I treat him quickly and re-bed the creep area with fresh straw. I will never let this happen again. The lives of these beasts depend on us.

I spend the morning cleaning out the houses. I muck and

pull and clean. I lime them to kill any infection and then throw fresh straw upon them. I muck out the hens too for good measure. They have not been cleaned out for some time, and cooped up as they are, they crave fresh bedding.

In my acts of work and cleaning, I find solace and peace and some perspective on my thoughts. I know that this struggle between father and son is playing out an age-old rural drama – these rows have been had by men like us for generations; we are the two bulls in the field sizing each other up. This is the way things are for now, but it will not always be such and the good days will come again and we shall talk of Charles Bronson and his harmonica.

I think at times I deserve this life, this hardship. For here now, in this winter that does not seem to end, it is hardship. The work is so relentless that I have forgotten I have lived other lives or that other lives exist. There is only the yard and cows and the mountain of chores before me. On the good days, I fancy myself the country farmer, Siegfried Sassoon reborn, the squire who will run an organic farm and make a difference in the world. On the bad, such as today, I think only of escape and leaving. But I cannot leave, for Ma has asked me to help them and I will not let her down. There will be deaths if I leave now, there are too many lives at stake. Mam and Dad have supported me for six months as I try to make it in this undefined world of literature. I must return that favour now and support them in the very real world of farming.

At midday I finish my chores. In the distance I hear the church bells call out across the fields from the village. I pause and reflect. Nothing has died, nothing has been broken that cannot mend. The sun too will return. I have learned to calm my thoughts, to calm negativity.

There was a time when that darkness had a powerful hold on me, but that was long ago, in another calving season. I do not think of that time any more, it is better not to. We do not talk of it either, save only in passing. I have come to love health and life and work, to think of each day as a gift, tomorrow a bounty, a land of unknown triumph and tragedy, and for all of it I am ready.

Fighting

The first record we have of man fighting the cow appears in the Mesopotamian *Epic of Gilgamesh*. In this poem, Gilgamesh and his friend Enkidu kill the Bull of Heaven, and the description sounds eerily like the bull fights of today: 'They fought for hours, until Gilgamesh lured the animal with his bright tunic and weapons and Enkidu thrust his sword deep into the bull's neck, and killed it.'

We know that the Romans fought against the aurochs, but it was the Spanish who really turned bull fighting into an art form during the time of the Moorish occupation in AD 711.

It was said that in the ring the Moorish and Christian knights could fight the bull rather than one another. This early bullfighting was associated with religious feast days. It was done from horseback with long spears or javelins, and thus was the picador born. The bull above all other animals fought the longest and bravest. It fought to the death.

It is interesting to note that, despite their Muslim faith, the Moors did not seem to mind the ritual slaughter of the bull, even though the deaths then were never in halal fashion. It is perhaps an example of how naturalized the conquerors became.

In medieval times, bullfighting was the preserve of the wealthy, but by the sixteenth century the nobility had released its stranglehold on the practice and modern bullfighting, the *corrida*, began. Now men began to face the bull on foot and with that came the figure of the matador.

From Hemingway to Picasso, many artists have been drawn to the fight and sacrifice of the bull and the man. In it we see still our struggle with nature, our desire to conquer it, and its resistance to that will.

Today's Spanish fighting bull is said to bear a very close resemblance to the auroch: tall, muscular and strong, its temper fearsome, its bravery renowned. I have seen the fight with my own eyes, in the Plaza de Toros de Las Ventas in Madrid. Locked in their dance, the matador and bull are a symbol of Spain – its machismo and pomp, its love of drama and, like its empire of old, its demise too.

I remember meeting José Manuel Mas, a young matador who had trained in this craft since the age of sixteen, who hoped one day to be a fighter of note, for there are fortunes to made at the sport.

I remember too the sounds of the band's paso doble and the entrance of the bull and the picadors. And when they had struck and injured the bull with their spears, I remember the arrival of the toreador.

The fighter stood forward and signaled to the picadors that he was ready. The crimson blood was flowing down the bull's neck, his head rolled and bowed but there was much fight left in him. It was a Miura bull, the most fearsome breed.

The banderilleros next came and stabbed the bull with their harpoon-shaped spears. Six harpoons were placed and the beast was weakened more.

The fighter stepped forward again and fought the beast himself now, armed only with his *muleta*, the cape, and his sword. The bull rushed towards him.

It passed once, then twice, and the man's turns were true and brave. He did not move from where he stood, his legs were taut and strong. The cape fluttered in the breeze, and up and over the bull's head each time as he passed, the red of its blood invisible against the red of the cloth.

The toreador performed a *rebolara*, a decorative pass, and the creature began to tire. I had heard that bullfighting was a ballet, and I saw then that it was not a dance of death, but rather a dance for and with it. In the end the fighter ran with his sword outstretched and the bull, as if complicit, dropped its head and bore the nape of its neck to the man. The act was clean and quick and the beast fell instantly. There was tragedy in the act, for I had never seen a cow so treated, to be raised so high and cheered by thousands, and then lowered to its death, dragged from the field by a team of horses. It was a moment from history. And a moment of history.

Across the road from the coliseum there was a café where you could eat the butchered meat of the fighting bulls. I was there with my Toronto partner, but she would not eat the meat and so we drank and smoked cigarettes like the tourists we were.

The fighting of the bull marked an important note in the history of the cow, for in this celebration of death was the celebration of the animal as life itself.

Texting

It has been five days and the red calf seems to be back to normal. We have survived through the bad week and Da has cooled and calmed again. On Friday evening, Ma and he go for their supper in town. They put on clean clothes and smile and joke.

'Have a good time,' I say and wave them off.

We have had a lamb born today and that has lifted all our spirits, for he was a triumph: he is big and powerful and will mature quickly. I will take the evening shift and watch the animals. It will be a quiet night, I think, so I can relax and watch a film or chat show. The grocery shopping has been done and the house is full of food. I snack on some peanut butter and lazily channel-hop. I rarely watch television any more, but tonight I find an old action movie and allow myself this treat. It is a Sylvester Stallone flick.

At eight there is a commercial break and I walk out to

the yard. A cow has been roaring on and off for an hour and something is wrong. A few days ago, a calf got stuck between the round bales of straw in the big shed. He was there for several hours before we noticed him gone. He was distressed and somewhat dehydrated but quickly recovered. I do not need a repeat of this, especially not after the row earlier this week.

I walk first to the bales at the top of the shed but find no calf. I return then to the roaring cow and trace her calls. Something is amiss. It is pitch dark outside and the lights of the shed are dim, and so I begin my search. She walks towards the creep and roars to the calves clearly now.

It's then that I find him. Stretched out, already cold, the life long gone from him. Red is dead. It is his mother roaring – no, not roaring, I tell myself, but crying, for she has lost her calf.

I make to curse but cannot find the words to express such disappointment.

The vet said you would live is all I can think as I drag his lifeless corpse through the shed. He was big and strong in life and so he is in death; my steps are laboured and slow, for I carry some four months of growth and life. We reach the outside and I release him to the concrete. I cover him in black silage wrap. It is so dark now that I cannot tell the plastic from the night. Tomorrow we will dump his body at the knackery.

I do not call Da, for I do not want to break his night off with Mam. They have both of them earned it. Instead I write a text.

'Red calf is dead, passed an hour ago.'

Da cannot text, though he can read them. There is no reply nor call back. He says nothing when he returns. There is no fight or row. But the calf is gone.

Days later, Mam tells me it was a bitter blow to him, but there is no one to blame. Death has taken its first victory on us. We will decide what to do with Red's mother in the days to come. She might raise another calf, but we have not the heart to think of that tonight.

Farmers will tell you that the loss of a calf does not bother them. That is a quite simply a lie.

Aftermath

It takes a few days to overcome the loss.

'Better outside the house than in,' Mam says, as we break for dinner.

She is right – it could have been Da, or herself, or indeed me in some accident. I do fear the day when there will be a call to say they are gone, and what then? What of all the said and unsaid things? It has happened twice already, with Uncle John and Uncle Mick. We thought it had happened when my brother got his arm caught in a machine at the factory and we feared for the worst, or that he might lose it, that he might never regain its use. But the surgeons were quick and saved him. He has metal rods keeping the arm together now, but he is alive and able bodied. Mam says it was the best thing that happened to him, for it made him slow down at work and settle down with his girlfriend and get married. They have a child together now.

I am aware now more than ever of mortality, for there is nothing like darkness to show one the light. It has come too in growing older and seeing death on the farm. We must be thankful that we have been spared.

The cow cried all night for Red, but come the morning she had stopped. Da disposed of his body. I offered to help but he wanted to do it himself. I read of St Luke recently: he has often been depicted as an ox with wings, and I think that perhaps Red is something like that now, in bovine heaven. I do not know where the spirit of life goes from a departed beast. I must ask Father Seán.

Da and I said nothing of the vet. Perhaps he missed something. Or perhaps medicine can only do so much and that is what he would tell us now if he were here.

Before the Department of Agriculture imposed stricter laws that required the removal of carcasses, we used to bury the smaller calves who did not make it in the fields. I remember their graves, if that is what they could or should be called. By the corner of the Garden field there are two, another in the yard where we keep the bales, and one in Mick's old potato ground. I was still a teenager when we last buried a calf on the farm, and the grave was big enough to hold me. I would wheel the body out in the wheelbarrow and begin to turn the sod with my spade. The ground in these parts can be wet and, though they are animals, I did not like to bury them in a watery grave and I always chose a dry patch of ground. The graves would be a few feet deep, not out of reverence but of necessity, for I did not want foxes or dogs to disturb the remains and so bring up disease.

Once the grave was dug, I would place the body into the ground, atop of which I placed a layer of plastic. I did not know why we did this, but it was something I saw Da do years ago, so I always did it too. I suppose it keeps the scent down, trapped in the ground. As a younger boy, sometimes I spoke to the calf and said a word or two, but as the years passed the words grew less. I lamented the loss, but I knew then that that was life: things died and you got on with the business in hand. There would be next season and the cow would hopefully breed again. As farmers, we must always look to the future, for the past holds nothing, neither feed nor money nor living.

There is but one fully grown cow buried in our fields. It was done out of respect, or *meas*, the older Irish word. The little Blue cow earned that grave. She was our first great cow, back twenty years ago now, when the farm was much smaller and we much poorer. She was a cross-bred Black Polly and Belgian Blue. She had no horns but a temper second to none. She was wild and unruly, but every year she gave us the greatest calf of the season. She was a thief too, and often led the cows on raiding missions through the parish for fresh grass. I remember once when the herd went missing and we spent two days looking for them. Their disappearance was broadcast on the local radio and neighbours came to help. We found them that weekend in Cullyfad Forest, several miles to the south. The Blue had led them all the way there, in search of what, I do not know.

The Blue died years later, peacefully, as an old cow. We had bought land beside the house from an elderly neighbour who was retiring. We found the Blue in the field as if resting, her body stiff, death upon her. We were not sad, for she had enjoyed a good life. She had done so much for us, as her calves had helped us buy this land. A family friend who was a vet said she had died of cancer. We buried her there to rest in the fields for ever. We talk of her sometimes still, fondly remembering her temper, her independent streak and her breakouts to fresh grass.

Cows are herd animals and in those herds there are hierarchies. The world of cattle is a female-driven one, for, as with elephants, there are dominant matriarchs who lead the pack. The bull is but an ornament, given to them for copulation and protection from other bulls. They are said to be his herd, but I have seen cows more fearsome than bulls. I do not know who took over when the Blue died, but there is a red Limousin in charge now. She is not the biggest nor the strongest, but she is the most fearsome. We have had to separate some cows from her, for she has bullied and beaten them. She is enjoying her throne now, but it will not last for ever; there will always be another rival, another to take her place. Sometimes I smile at this game they play and think it is like a bovine *Game of Thrones* out there in the fields. We even have our own dwarf in Napoleon the calf!

On Wednesday, unknown to us, a cow gave birth. The new calf is grey and his face is a little like Red's, though he is more muscular. He is sucking by himself and does not need me. Such is life and luck.

Film

Father Seán and I sometimes go to the cinema together. A few weeks back, we went to see *The Revenant*, which had taken some time to come to our local movie house. It was a powerful and epic story. Father Seán covered his eyes at the gore. It was strange to behold, for he sees death – real death – every day in his parishioners. I know of no one else, save a doctor or nurse, who has been surrounded by death and illness so long. He does not fear it, but he does not like the gore.

We share our popcorn and when the movie is done we agree that it was profound and moving. We talked then for a long time of the Native Americans. Father Seán has a great affinity with those people and the frontier of America. He has read many books on those times and can tell me of the Apache and the Sioux, of Sitting Bull and Wounded Knee, of Chief Joseph, whom he calls a great man, a visionary.

On taking the leadership of his tribe, the Wallowa band of the Nez Perce, Chief Joseph, or Hin-mah-too-yah-lat-kekt (Thunder Rolling from the Mountain), never expected that he would do battle with the very United States itself. But then, as was their wont, the Yankees forced the tribe off their lands, which Joseph had sworn to his father not to abandon.

The resulting war occurred soon after the defeat of Custer and made headlines across the world.

Chief Joseph was a deeply religious man and I think now that it is the spiritual aspect of him which appeals to Father Seán.

In a speech at Lincoln Hall in Washington DC in 1879, Chief Joseph said:

> Our fathers gave us many laws, which they have learned from their fathers . . . They told us to treat all men as they treated us; that we should never be the first to break a bargain; that it was a disgrace to tell a lie; that we should speak only the truth . . . We were taught to believe that the Great Spirit sees and hears everything, and that he never forgets . . . This I believe and all my people believe the same.

It is an old story and an old war, but Father Seán denigrates neither the man nor his cause. He has too much respect for those great peoples. Chief Joseph and the Nez Perce lived out their days in exile, and by 1877 for every Native American in the West there were nearly forty whites.

Father Seán reminds me that Ireland cannot forget the Native American people, for they sent us aid in the Great Famine.

The film and talk over, we make our way home. Father Seán jokes to me as we leave the cinema that perhaps people will think us a gay couple, for we have been to see two movies together in the last few weeks.

'Let them talk,' I say. 'They have little to be at.'

He relaxes at my ease and agrees that I am right. We are unlikely friends, for he is seventy and I not quite thirty. He too is from a farm and knows cattle. I think my farming sojourn has reminded him of his own youth, for between our talk of books and the Apache, he has unfurled the stories of his past, of the triumphs and losses of calving seasons in the long ago. Once he delivered a calf while in his priest's outfit, and someone joked later that it must surely be God's own work.

He has not lost his deep love for nature and often goes on long country walks, where he takes pictures or paints watercolours. He has shown them to me, and he has an artist's eye. He was once an architect before becoming a man of the cloth. He was once in love before becoming a priest. He has lived a full life.

Father Seán's family have been in Longford for over a thousand years, as keepers of the church land in his native Killashee. His is an old faith.

Our movie nights break the routine of the farm, and do me good.

Accidents

I've a sore foot and I thought a few days' rest from running would cure it, but when a hungry ewe stood on me yesterday and I promptly spent several minutes hopping about and swearing for the pain she had caused me, I knew something else was up.

I've asked around and have been told to go to a local physio. She is a no-fuss woman and puts me upon her treatment table. She enquires what my profession is and I tell her 'farmer'.

She inspects my foot, pushes the tendons this way and that, then looks up and informs me that I have damaged the ligament.

'It was either your wellingtons or you hit it running.'

'I thought it would heal itself,' I say.

'No, hasn't a hope in hell, but I'll fix it right up,' she says. 'You don't mind needles, do you?'

I shake my head. 'Whatever it takes, I want to get back to the farm.'

'That'll do. You might want to tell me to fuck off in a few minutes and that'll be OK too.'

'I wouldn't do that,' I say.

She instructs me to lie flat and begins to dry-needle my foot. On her wall is a chart of acupuncture points; the parts of the body are written in Chinese and English. I study this map as she prods and pins and I feel a calm come over me. There is a great surety in her hands, an understanding of the body and how it works. I think of my own work with the lambs and how I know what I must do with each of them and how I am gradually coming to understand the subtleties of the cows too. It takes years to learn their ways. It is a special knowledge to have, for not all possess it.

'Fuck!' I yell suddenly, as she locates the pain and the needle scratches what seems my very bone.

'Now, didn't I tell ya?' she says.

'You did.' I laugh.

The process takes a few more minutes. I begin to feel the pressure lift and it seems that the pain does not leave but flows away. We talk now of farming and football, of Australia and Canada, and of Uncle Davy, for she knows him well. It seems there is no one in the county who does not know him. She fixed his back, she tells me.

'The Connell men are prone to a bad back,' she says.

'Well, it's funny you should say that,' I begin. 'I've been getting a twinge.'

'Well, roll over there, and we'll take a look.'

'My foot?'

'It's done. I finished when you were chatting there. It will be fine in a few days.'

I roll over and take off my top and she runs her fingers along my lower back.

She prods and feels and informs me I have a popped pelvis and that it's been that way for some time.

'Begod, and what caused that?'

'Well, farming is hard work. You've been doing a lot of lifting and repetitive movement, no doubt?'

'I have.'

'Well, that would be it. This is going to hurt. You can curse again if you want,' she says.

I swear and curse loudly now as she rights me. I stand upright and walk straighter now. I am mended and feel clearer. She does not overcharge me and I can afford the fee. I have not earned much money in the last few months, for the farm has been my job, but what little bits of writing or journalism I have done have kept me afloat and I have not starved or wanted.

We bid each other well and I thank her, for my foot feels so much better already. I know now how the cattle feel when they are unwell and I fix them. I smile and think how simple that was, and feel foolish for not having gone sooner. I think then too that this is so small a matter in the great scheme, for I heard only a few days ago that a man had died in a farm accident.

Farming is the most dangerous job in Ireland. There have been 200 deaths in the past decade. Some have been horrible, some careless, and others heart-breaking. It is a small island we live upon, and each death is known throughout the country. Behind every statistic there is an individual, there is a story.

There was an incident, some years ago now, in which a family lost a father and two sons in a slurry pit. It was spring and the tank was to be emptied. To do this, you must first agitate the manure using a giant mechanical blender, which liquefies the shit, making it easier for the slurry tanker to suck it up and spread it on the fields.

Agitation is a dangerous process, for cow dung is full of methane and the gas released can cause blackouts and dizziness. It must be done in a well-ventilated shed, and you always need a second man present in case the fumes consume the operator.

A dog caused this accident, for it somehow fell into the tank. It was near the end of the day and there were but a few feet of dung left to be spread. The farmer had done everything right, but, on seeing the dog stuck, he forgot himself and went to rescue it. I suppose he thought that with so little left in the tank it would be safe. But the fumes knocked him out and he fell into the manure. His son next went to save the father and it took him also, and the last of the brothers died trying vainly to save his father and sibling. They were all of them drowned in a few inches of slurry. The dog alone survived.

The priests prayed for that family at mass. We mourned their loss and thought how easily it could have been us. It is only in death that we remember our transience, our fragility. We are but passing through this world and passing through these farms.

Out with the Old

Some of the cows are getting old and worn out and must be replaced. One or two have not gone into calf this year and have been fallow. Sometimes this happens and if it is a good cow, we allow her the rest year, but these two ladies are old now and the calves they gave us each year were only ever average, so it is time to let them off.

A new star system has been introduced by the Department of Agriculture which rates all cattle out of five in an effort to raise the blood lines of the national herd. Farmers must now aim to have a majority five-star herd. So far there has been a lot of resistance to this new system, for farmers have spent generations breeding up their herds to a point where they are happy. The new system does not favour self-breeding. Rory, our neighbour, calls it a monopoly by the genetics companies to impose their sperm banks on farmers. Perhaps there is truth in that too.

We tell ourselves that the purebred bull calf will surely be a five star, though we have not yet traced his breeding to find out. He is growing well and it has been agreed we shall let him and his brother out to grass, now that the weather has calmed somewhat and there has been some growth. It is healthier for them outside, with less chance of diseases. We will put them in a small paddock and bring them nuts each day and keep a good eye on them. The purebred calf is like a baby still, and he must be minded and cared for.

The remaining four weanlings who were too young to sell can also go outside now, along with two of the cows. We will move them to our hill farm of Clonfin, which was once part of the Thompson estate. With its fields lined with beech and oak, it has the touch of England about it. The house, which still stands, is from the 1700s and belonged to the game keeper. It was last occupied twenty years ago, by an old relation, Dolly. She is dead now and her line is gone, for she was a spinster, and the home is now a ruin.

Dolly lived a frugal life, without running water or electricity, and her world was unchanged from that of her parents in the late nineteenth century. Her father was killed by the English in the War of Independence in reprisal for the Clonfin Ambush in February 1921, where Seán Mac

Eoin, our local hero and commander of the Irish Republican Army, led an attack on the occupying forces. We won the day, killing many of their men. Dolly's father was too old to fight but not old enough to die. They smashed his face open with a rifle butt. It must have been a horrible end. Afterwards, it was said that bad luck flowed upon the ground from that act.

Moving the cattle takes most of the morning, but finally the weanlings are settled in Clonfin, and the purebred and his family are placed in the small Garden field by my brother's house, over the road. Because our farm is spread over many parishes, we have to move the cows to different locations according to the season. But when the cows arrive back in a field they haven't seen for a year, we have never to show them the water drinkers or springs or hidden places of shade and rest; they remember it all. Thinking thus, I wonder: do the cows remember their losses, the crops of calves we take from them each year? I do not know.

Thanks to the poor weather, the field isn't nearly as lush as it was last year. At least there is grass in the ground now, not so much to cheer about, but it will feed them for a time. Anything over ten degrees will make grass grow, so we must pray for warmer days.

Returning home, we prepare the two old cows for their departure. Da will do this business, for he knows and likes the mart. He never asks me to come, though I think I am getting to be a good judge of cattle now.

'One good heifer would make up for them both,' he says, as we load up the trailer.

I wish him luck and safety.

He does as he says and brings back a cow in calf. She will calve in three weeks. He tells me she was a fair price. I give her a house to herself for the night. I will move her to join the rest of the herd in a few days when she has settled.

His Master's Voice

Every farm and every family have their own unique calls for cattle. These noises are a form of oral culture passed down father to son. The cows know this language and newcomers to the herd quickly learn it, so that they all understand what the words or cadences mean and respond when we want to move them. So often, the words are not words at all, they are not English nor Gaelic, they are of an older sort of sound, perhaps from before, from the long ago.

I have read of the Fulani people of Africa, who are the largest group of nomadic herdsmen in the world, numbering some 13 million people. They still adhere to their traditional way of life, moving their animals across the plains of Central Africa through the seasons. I would love to hear their calls, for they must be very old and ancient, unchanged for centuries.

In Australia, where the farms are vast, I have seen dogs used to herd the cattle. They jump on the back of a quad and travel with the farmer out to the bush. The blue heeler is a powerfully built animal, with great personality and tenacity. It will bite cows on the nose and hush them forwards. Here, though, things are different. Our cows do not fear the dog and will stand and fight him. Vinny is young and, with that, foolish, but he knows enough not to go up against the bigger cows. He is, after all, a sheep dog, and it is not in his nature or instinct to move cattle.

And so with sticks and wire and calling we move the herd. There is a psychology to this act: one must predict what the cows will do when certain calls are made. To gather them to us, we shout, 'Suck, suck, sucky,' which grows faster, eventually flowing into a continuous rolling *s* sound. This calling may take several minutes if the herd is in fresh grass, and sometimes it does not work, for, being sentient creatures, cows have their own free will.

Family calls can sometimes change. I learned a wolf call from old Robin Redbreast years ago and now it is part of our farm's vocabulary. It mimics the sound of a wild dog, and it has never failed to move a cow or sheep forward. For though neither the cows nor I have ever seen or heard a wolf, the noise is buried in their DNA, in their instinct, and they fear it.

When the pack is moving, we yell and keep up our shouts. Hup, hup, hup, ya, ya, ya, heyup. These are old words, words that were used on the ox and working horses by Granddad and Great-Grandda. With them, we sing the cows into the crush or holding pens. We picture them in our minds standing in the holding pen, and the song is the vehicle for that vision. This reminds me of the Aboriginal people of Australia, who sang their country into existence through their songlines. After all, as Bruce Chatwin wrote, the first words began as song.

The cows answer us with moos and calls. We try and not run them, for one may break away from the herd and cross ditches and perhaps break into a neighbour's field, and so, as we move closer to our goal, our calls grow softer and we tell them then that they are good girls. We cluck and coo and they calm and slow, and in the end we sooth them with gentle shushes. In this they know the annoyance is nearly over. Sometimes on hot days when they are in the crush, we scratch their backs with our sticks; they enjoy this and keep calm and docile.

It is how we speak, they and I, and yet there have been times when I have spent many days straight with these creatures and have wished we could communicate properly. When the Tower of Babel fell, it not only divided man but species too.

Sticks

What with all the moving of livestock, it has become clear that we are low on herding sticks. We sometimes use rubber piping or plastic rods, but they soon break or bend. It is my job to get new sticks each season. I have collected them every winter for the last five years. I only take from the trees what we will use.

Today I bring Vinny with me to do this job. It will be exercise for him and good training. He must learn the different fields and farms and how to negotiate them. One day he will be grown and working alongside us. He must be taught how to work, as I was.

The hill farm Clonfin has the best plantings. Herding sticks must be straight and are best taken from a growing tree and left to harden. A good stick might last years. It is an investment in time to find the right ones.

There is a place at the bottom of Clonfin where the land meets the nearby bog. It is a wilderness now, for the bog is no longer used and the whole area has returned to nature. We have dubbed it the 'Wild Wood', from the children's story *The Wind in the Willows.* When Javine, my sister, was a child, we told her this is where Ratty and Mole and Badger lived. When she was naughty, she was threatened with the weasels who would steal her away, as they had done with Mr Toad. Though a teenager now, she still remembers the weasels and the wood. The place has always held great magic for her. It feels old, older than this land. This is where I harvest my herding sticks.

Vinny and I travel up together in the jeep. He has learned to jump in and out of the vehicle now. He is not afraid of driving any more and he no longer defecates in the back. I talk to him as we drive and tell him it's fine and we are nearly there.

The cows and the four weanlings are on the back hill when we arrive. They are busy eating and do not bother looking up, perhaps sensing that we have come for another purpose.

The ground is still wet and Vinny's paws are soon mucked and dirty. He jumps on me from time to time and I shout to him to sit down, for I do not want to be tracked in mud.

We walk towards the Wild Wood and a pigeon bursts from the ditch. It flutters through the air and comes to land on the other side of the field. I photograph it in my mind's eye. There are hawks here too, and owls. Once, months ago, I saw a great battle take place here between a hen harrier and pheasant. The pheasant, though agile, was soon killed.

The Golden Eagle Trust has done great work to help Ireland's birds of prey in recent years, and more and more great hunters have emerged once again, including buzzards, falcons and kestrels. Sadly not everyone has welcomed these creatures, and some men of my profession lay poison bait for them, fearing that they will take lambs and fowl from their farms. These fears are not unfounded, for I have seen a buzzard kill turkeys that my aunt was rearing. The bird was shot in the end. It hurt me greatly to see the great creature laid low but, as my aunt said, what could be done?

There is just one tree I harvest each year. I have not told Da or the others about it, for I do not want it to be over used. By the brow of a ditch it stands waiting. It is an ash tree. The wood of the ash is good and its saplings have grown straight and hard. They are perhaps seven or eight years old. They are just the right width for a man's hand. A sapling too big will not make for a good stick, for it will not have good balance and will not sing through the air. I have brought with me a small saw and begin to cut the saplings. Each stick needs to be around a metre and a half long. There is no give in this wood, which means that when it strikes the cow she shall feel it. But it is foolish to overuse the stick, for you will scare the animal, and a worried beast is a dangerous one.

The ash tree is a magic tree and one of the seven mystical ancient trees of Ireland. In older times, it was used for spears and weapon handles. It is also known as the World Tree, for it was said to be the route from the underworld to the very heavens above. Everything in this country has a story. Everything is rich with meaning.

Vinny plays in the undergrowth as I finish my chore. I am careful to cut the sticks fully. I want a clean cut; to simply snap or break off the end will fray and wear the timber unnecessarily. I take six sticks. Two will be for Da and me, another two for extra help from Mam or my brother, and two spares, for they may be taken to a mart or another farm and forgotten. That should see us through the year.

There is a holy mountain in the west of the country and it too has a connection with sticks. The last Sunday in July is Reek Sunday (or Garland Sunday), a day when tens of thousands ascend Croagh Patrick, Ireland's holiest mountain, where it is said Saint Patrick fasted for forty days and did battle with the devil. It is an ancient place and in older times we worshipped other gods upon it. On the Reek day, the gypsies sell and rent sticks to pilgrims; 2 euro will rent you a stick up and down the mountain, and 3 euro will buy you one to keep. The tinkers' sticks are always good and straight. The climb takes many hours and some do it barefoot. A few years ago, I myself felt the need to climb it. I was looking for a connection with the mystical, which is why so many climb. There are people who will tell you that when they reach the summit, bloodied and beaten, and look out on Clew Bay with its hundreds of islands, they have found it. But I think I was looking for answers that no mountain could bring.

My sticks gathered, I call for Vinny and we make our way back down the hill. There are lots of saplings left for many years to come.

When I get home, I sand and plane the sticks. I am slow and careful, for I do not want any burs or sharp edges to cut our hands or to hurt the cows unnecessarily. I stack them and leave them to dry out for a few weeks in the shed. We shall use them in the spring. I let Da know that the sticks will be ready.

'We were running low,' he says.

Gadarene Swine

I no longer eat pork – I have abstained from the flesh of this animal for nearly a year now.

The breeding of pigs for meat is one area in which farming has broken fully with nature. I toured a piggery a few years back.

From birth, the pig knows only the piggery. Born in a litter of ten or more, the piglets will suckle for several weeks, while their mother lies on her side, caged in a farrowing crate. This is done to prevent the sow from eating her young, which can sometimes happen. The piglets are then removed and taken to a separate weaning room. There they join all the other litters to be fed through feeders. The pigs can eat as much as they like, they simply push a small button with their snout and a watery feed mixture emerges. It is kept very hot in these rooms to promote fast growth.

The farmer must be careful to prevent bacterial infection, for in these sterilized environments bacteria will spread rapidly. Deaths are common at this stage.

Walking through the piggery that morning, I saw the workers pull the dead bodies out of the rooms, slinging them to the floor.

The floor in the piggery is slatted; there are huge tanks underneath which collect the watery shit and this in turn is sold as a fertilizer to farmers. Its smell is second only to that of human waste, for the pig is an omnivore too.

The piglets who survive are brought to a new house, where the best females will be selected as breeders; they are the lucky ones and shall get to live. The males are taken to their own section and there will be fattened.

The pigs here tend to be bored and a bored pig will sometimes bite the tail off his fellow. In the past the animals' tails were simply docked but now due to changes in EU law many farmers provide straw and devices to reduce animal stress. In cases where tail biting still occurs farmers are allowed to dock tails. The widespread practice however still occurs in other countries including the US.

Both boars and sows will be fed until they reach the right fat-to-meat ratio. The weights are set by the slaughterhouses: the carcass should kill out at around eighty-five kilos. If pigs are over their weight, some factories will charge penalty rates, so the farmers keep a close watch on the weight of their beasts.

In Ireland, males are killed before sexual maturity, for this prevents boar taint. Boar taint is a bad odour that emerges from the meat when cooked, caused by increased amounts of testosterone in the beast's body. Pre-maturity boars make for the best and softest meat. As a result, Irish sausages are known across Europe for their flavour and unique taste.

The pig industry is controlled by just a few farmers in Ireland. Many of these farmers have built empires on the back of the swine and run many such factory farms. It may seem a cruel industry, but it provides employment in the local areas, and from these small factories the nation is fed. Despite it all pig farmers do care for their animals. I think however few people know where their pork comes from. Fewer still know how the animal might have lived there.

It is the way of modern intensive farming to remove nature from the process as far as possible. There are around 1.5 million pigs in Ireland, and few of them will have ever set foot on grass or known the feel of muck upon their snouts.

It has been over a year now since I last ate pork. It was a decision born of both the spiritual and the humane. I miss it though, I will not lie; I miss rashers, I miss black pudding, but the pig is one of the smartest animals in the world and knowing how they live in order to produce these items no longer made them taste good. It is a small sacrifice. Maybe my Jewish and Muslim brothers are right.

Risk

The ewes and lambs we rehoused a few weeks ago because of the bad weather have contracted orf. This viral infection is highly contagious and spreads quickly from lamb to lamb. It causes sores and scabs to develop on the animal's mouth and face, and those lambs badly affected can find it hard to suckle their mothers due to the pain of the sores. In the worst cases, they cannot suckle at all and starve to death.

There is a vaccination for the infection, but it is too late now to have an effect. The disease will pass of its own accord in a few weeks and the affected animals shall have a lifelong immunity to it. But knowing the course that the disease will run does not stop farmers from worrying, and a healthy faith-healing business has emerged. There is a woman with a cure for orf and, driven as we are to try and treat the condition, we call her now. She asks only that we hold the phone up so that she hear the animals over the line, she will then say a special prayer and the orf will clear. Gormley the vet does not hold with such beliefs.

Uncle Davy's sheep have come down with orf too, so I suspect we have infected each other's herds with our travelling up and down. We know it is as a result of the animals being housed for so long. Though I have bedded their shed with fresh straw, there are over ninety animals in the space. They need to be outside, where it is clean and fresh and infection cannot take hold. It is but another lesson that nature is best.

Time has moved on and the days become weeks; it is supposed to be nearly spring but the weather is still not much better; it is however agreed that we will let the animals out once more.

'They're better out than in,' Da says, as we drive them towards the upper ground.

'Aye.'

There are over a hundred sheep in the field now. It is more than we have ever had. I do not know if the ground can take so many, for they never stop eating.

'We'll bring them up nuts and I think we could start giving the lambs pellets,' Da says.

'We could. And the silage?'

'We'll keep up with a little bit to them,' he says.

'It's no harm.'

We agree that we will feed them some silage each morning on our rounds. It would be better if it were hay, but we have to keep the hay for the newly calved cows. It is a rare commodity.

The lambs are happy to be released again and, though it is wet, they jump and prance as we walk them up the lane. Once everything is in place and settled, we load up the tractor with the feeders and nut dispenser.

We shall give the mothers a bag full of nuts each morning, since there is not the goodness in the grass yet.

The lambs have different nuts, which we put into the nut dispenser, which is known as a creep feeder. This holds several bags of lamb nuts and is a type of small shed, which only the lambs and not their mothers can fit into. The lambs soon learn the purpose of the feeder without my needing to show them. Despite the bad winter, they are growing well.

Even after three seasons with the sheep, I am amazed by just how quickly they develop and grow. The calf's maturity is a slower thing.

Vinny has joined us on our jobs this morning and stays close by me. He is still afraid of Da, or perhaps does not know him so well. There was a time the dog would have chased the sheep, but not any more.

At twelve, I begin the clean-out of the shed. I must clear the straw from it, for the orf might well be alive in the bedding and infect the next group of lambs and mothers. I muck and pull and drag the bedding onto the front loader and after three hours the job is done. I will not need the gym today. I disinfect and lime the area thoroughly. I want to kill the badness.

Dolmen

Sometimes I go and visit the dolmen in a nearby village. It is an ancient megalithic burial tomb, consisting of three standing stones and a large flat capstone perched atop. It is older than the pyramids of Egypt or Stonehenge, and yet here it stands in a quiet field behind the village of Aughnacliffe. Few outsiders even know of its existence. We are a superstitious people still and would never harm it, for these strange stone structures are said to be portals to the other worlds: the land of youth, the underworld, and beyond.

Ireland has a Valley of the Kings like the one in Egypt,

though it is not so famous. In the Boyne Valley, in the nearby county of Meath, our ancient high kings ruled and were buried. I have seen it by daylight and moonlight and it is a thing of beauty.

My wellingtons are wet with dew and a cold breeze flows in from the brow of the surroundings hills. I breathe in the air and linger, looking through the stones' arch, looking in the hope that there might be something else visible on the other side. As a child, my heroes of mythology were both Spiderman and Cúchulainn. The feats of old Celtic gods and heroes intermixed with those of American comic books. I have left Spiderman and the X-men in childhood, in the distant past, but the works of the old world, the Celtic world, fill me still with awe.

As hard as I look though the dolmen's portal, though, I see only the green fields on the other side, but they are a vision unto themselves, a vision of my natural inheritance, and with that thought it is time to go. Some cattle graze upon the nearby grass and I think now this is a picture of life from long ago. Back then, a farmer grazed these lands as I do now, his cattle beside him, paying reverence to the old ones.

I have made it my business to bring friends and lovers here and show them this place. And I shall keep visiting the dolmen, even though I do not fully know what compels me to come.

Cycling

I got a bike at Christmas but the weather has been so bad that I have rarely been out on it. Today is somewhat better, and so this afternoon I decide to go for a ride. I take in the parish and cycle by our neighbours and friends. There are few cows or livestock out at the moment for want of grass.

There is a charity cycle to take place in a few weeks in memory of a neighbour who died suddenly. He was a wonderful community man and a great champion of our local football team, and the parish has mourned him deeply. I've yet to decide if I will take part, but I hope I will be fit enough for the fifty-kilometre circuit. It would be nice to race with some men my own age, for I have been so busy on the farm I rarely meet young people around here any more.

I do get time to talk to friends in other countries through my phone – friends living other lives. Tim and I talk every few days now. He too is a country boy, but a musician who is trying to make his way in this creative life. We compare the music and writing business and find it similarly strange. His music is fun and vibrant, it has the sound of success; he dreams of playing Glastonbury as I dream of speaking to large auditoriums. We have shared many losses and triumphs at long distance. We laugh and joke and it breaks our work and week. When the season is ended, I have promised to visit him in Spain.

I am glad I have friends from elsewhere, for the countryside can be a lonely place. I think that is why the pub holds such a position of power in rural Ireland, for it is the café of our world, the meeting point for talk and exchange of ideas. But I do not drink any more and so do not go. I don't miss the drink, for it never did suit me: I said too much under its influence and pushed others away. I prefer my way of living now to the barstool chatter of before. A long run is as good as ten pints to me. I often tell this to Da, which makes him laugh.

I cycle on and begin the slow ascent of Cairn Hill. This is our local mountain and it is said that Queen Medb's nephew and murderer is buried atop it. It is the highest point in the midlands and has defined and shaped my life in these parts, for it has watched over all our lives and seen the changes that have taken place. When my brother was a boy, he would tell Mam and Dad not of an imaginary friend but of his imaginary cattle upon its brow.

The hill is long and slow and my legs grow tired, but I do not stop. It will get easier, I tell myself. In the weeks to come I will cycle and run harder and I will get stronger.

I reach the top of the hill and stop for breath at the McCormack's. They have the highest house in the parish and are good farmers too. A lovely white heifer moos from across the ditch at me. I smile to think what must she make of me in my Lycra tights and cycling shorts, then turn and race down the hill for home.

Signs of Life

Jonathan, our scanner, is coming to ultrasound the cows today. We have four animals to check to see if the artificial insemination has worked. If all goes well, we will have four pregnant cows this afternoon.

Jonathan is from Cavan, the county to the north of ours. He is a nice man and when we don't talk of cows, we discuss fishing. Trout are his favourite catch. He is a great aficionado and competed in the world championship last year. The rivers here have much life in them. Da told me once that in his boyhood Granddad caught a wild salmon with a pitchfork and rope as she came up to spawn in the small river by the house. It's been years since we have seen salmon there, but it is still rich with trout.

Da and I walk the four cows down to the crush in the lower shed and load them into the shoot. They are all of them big girls and it is a squeeze to fit them all in. With much shouting and coaxing, they agree and enter. The crush was one of the first things Da built in the yard. It is twenty or more years old and is beginning to show signs of wear. We say each year that we will rip it out and replace it with a better one, but it is still here.

A good cattle crush is an important thing on a farm, for it is where certain medicines are administered, calves are born and TB testing occurs. We must test the cows for TB each year, as a requirement and condition of the Department of Agriculture. The vet must come to test them each summer and should a cow be found to carry the strain, she will be put down, and in extreme cases a whole herd may have to be slaughtered, so seriously is the disease treated.

Gladly we have not had this happen yet. We sigh relief after each year's tests, for to destroy this herd would finish Da and Mam. It is a culmination of years of buying and selling and breeding. There are some who say badgers carry the bacteria, but after years of eradication programmes the badger is not dead, nor is the disease. Indeed, many now question this link with the badger and in many places this gentle and misunderstood creature is at last being left in peace.

There are two Reds, a Simmental and a Black Whitehead in the crush today. The bull has been with the Simmental and the rest were covered with AI by Uncle Paul, so we shall see what has worked best. Da is keeping the bull away from all the cows now and has begun to make trips with our neighbour Rory to inspect potential new stock bulls. As long as the beast breeds well and is not temperamental, then I shall be happy.

Our very first bull is still the best we have ever had. We called him the Master; he was a Charolais from Ballinamuck and he was big and strong and kind. The farm was expanding then and we needed a bull to cover all the cows. It was, I suppose, a milestone in the growth of the place, an earmark of success. The Master was with us for several years and we had never any trouble.

At first we kept a chain attached to his nose-ring to slow him down lest he harm us or the cows. But that first summer, in Clonfin, he had been chasing two cows in heat and his chain became tangled in a tree stump and, pull as he might, he could not dislodge it. I must have been fourteen or fifteen years old and though he pranced and pawed the ground at the sight of me, after a time he calmed, for he knew I was there to release him. I remember it still, for he bowed his head low and became still and then I gently undid the chain. I did not speak and he did not low. Another animal might well have charged me then, but he did not: he simply slowly stood up and walked away and after a time resumed his hunt for the in-heat cows.

In that moment, it seems to me that he and I had shared a look across the gap of species. This is something that John Berger talked of in his work *About Looking*. Berger's words have shaped how I view animals; through his prose I came to appreciate a quality in their gaze, a look that we share each and every day on the farm. Indeed, his writings on animals were the first beautiful words I ever read of farming and beasts. Reading them then as a young college student eager to abandon the countryside, I saw that there was perhaps something more meaningful in nature and peasant life than I had first thought, though it would be many years before I would come to know that truth in my heart.

I knew the measure of the Master that day and I knew the measure of my own ability with cattle, for I had been afraid as I approached him, but I could not leave him to suffer, and perhaps he knew and respected that. We have never named another bull since the Master, but perhaps the new bull will earn a title. We shall see what Da brings home.

Jonathan puts on his overalls and hands me his ultrasound.

It is a small device which slips inside the cow and scans her like a woman with child. The machine has a small screen and through the grey and white images he can see what is life and what is not. He has never been wrong. Not in my time.

'Now, we'll make a start, men. How far along is she, Tom?' he asks.

'I'm thinking a few weeks,' Da replies.

'You're right, there. Six or seven weeks.'

'Great,' Da says.

They move on to the next cow.

'Now, she was with the bull,' Da explains, 'but I haven't seen her a-bulling for a long while. She's a tricky one, though.'

'She's a good cow, good meat on her,' says Jonathan.

'Well, if she's not in calf, it's the mart for her.'

Da has said this before. He hopes that she is in calf, for she is a good cow, but we cannot afford to have too many fallow cows. She will be another man's problem or another man's diner.

'Four months, Tom.'

'Begod, the sneaky yoke.'

All but the Black Whitehead are in calf. The job has been a success. She will be AIed again, but that is OK, for it is early in the season and there is still time for repeats.

There are some who say it is a cruel practice to subject the cows to year after year of pregnancy and calving, but they have been bred for this purpose. Indeed, it is the same in nature: a cow will calve each year until she can no more, and then an old matriarch will be killed off by predators. The world wants beef, and so long as the animals are treated with respect and the calves cared for, there is little to complain of.

Jonathan and Da settle up the bill. Jonathan will be back in a few weeks to check the next few girls. His appearance is always a breath of fresh air and today he has brought good news.

Doherty

We lost Mickey Doherty four years ago now. There is not a month that we do not mention him. He was Da's best friend and our neighbour. He was also a senator in the Irish parliament and an advisor to former Prime Minister Albert Reynolds. He was an important man in the area and Da was the closest thing he had to a son.

Doherty never married and was a member of the bachelor class. It was not such an odd thing then, for so many men of his generation never eloped. Along with Doherty, there were the Wild Men of Soran, a pair of bachelor farmer brothers, the Scanlons and the two Neds. They are all of them gone now. In a way, Da is the keeper of their memories and stories, and with it our link to an older time and an older Ireland.

Doherty was like a grandfather to us children. He was also the lynchpin of our neighbourhood and gathered us all to him on special occasions. My memories of our native sports of football and hurling shall for ever be entwined with his house; for it was not to Croke Park, the great stadium in Dublin, that we travelled on match day, but to his small kitchen. There we saw the trials and triumphs of men and teams. We shouted in that amphitheatre of brick and mortar and cried when our men were beaten and broken. At half time, he prepared tea and brack cake. It was his custom and in its repetition it became our ritual.

In his last year as his mind began to wander, Doherty grew closer to nature, and whilst he was unable to recall dates and events, he began to commune with the birds of the air. Senility is best described in the old tongue, *duine le Dia*, for in that phrase is a kinder, more understanding view of the condition. Its literal meaning is 'a person of God', for only their maker now can understand them. Perhaps in losing part of himself, he gained something more, for there had always been swifts and swallows at his house each summer, but in those final months they became his friends. Often I saw them enter his open door and fly around the kitchen. He greeted them and they in return sang out to him. He allowed them to build a small nest in the porch and there they raised two crops of chicks that year.

I remember too his last night in that place. It was a summer's evening and our cows had no water, for their drinker was broken. I do not know why, but I decided to walk rather than drive to them, and brought with me a book to read as I waited for the water to fill in the blue barrel. The chore took some half an hour and the evening was bright and clear. When the sun is with us, there is no place quite so beautiful as Ireland, with its dandelion-strewn meadows and thistledown gardens, its stubble-mowed fields and its long, bright nights.

The cows lowed and drank as quickly as the water filled, and after all had supped and the barrel was full I rose and left.

That same voice which had instructed me to walk now told me to stop into Doherty on my way home. I found him panicked, and he began uttering the mantra that played over and over again that night.

'I'm not right, I'm not right.'

'Should I call someone?'

'I need someone, I'm not right.'

On and on this talk went, until I called our neighbours Christine and Michael Lee and together the three of us decided what must be done. An ambulance was called and Doherty was wheeled out of the house. He never set foot in there again.

Months later, I sat in the nursing home with Da as he daubed his second father's dry, cracked lips. The dying man now gasped for air in a slow, rattling rhythm. My father did not speak that whole night. His movements said more than any words I could muster then or can write now, for in his act was truest love and emotion. Doherty passed a few hours later, and so ended an era of our lives.

Doherty's is sold now and new people have moved in. They repainted the place and have made it shine once more; there is a beauty to it that it has not known in years. They are quiet neighbours and we see little of them, but it has been agreed they are fine people. Granny says all that's missing is

Doherty in the doorway, his keys hanging from the latch and the swifts playing in the breeze.

The Gun Show

There's talk that our purebred calf will be a prize winner. His points and form are right, he is tall and though he failed for a week or two after we burned his horns, he is back on the mend. Rory is urging Da to begin his training, for it will take time, but Da is not so excitable and is happy to wait.

The showing of cattle is a particular craft. I do not know enough about show dogs, but I imagine it is something similar. It is an insular world that does not break ranks, a world full of characters and pomp, and sometimes villainy and pettiness too.

It has been a long time since we have shown a bull, and the talk has brought back memories of Eric and Envoy, our Simmental show bulls. I remember Eric especially. I was eight or so at that time and the training of Eric had been going on for several months. When he was a calf, his nose was pierced and a ring placed within it, next he had a halter placed around his head. He wore this for two weeks straight, in order that he might come to understand that it brought him no harm. The breaking of a bull is a slow process, for it has not the intelligence of a horse and so one must be patient.

When Eric had grown used to the halter, we then tethered him up to a wall. This act alone is what determines a good bull or not, for if he cannot settle and accept the tying, he shall never be trained. We started small at first, tying him for only ten minutes or more. Fight as he did, he was not strong enough to pull down the wall; in that he learned that the rope was a powerful thing and that only we could relieve him from its strange power. It took several weeks to build up his tolerance of being tied and to educate him in its way.

To lead a bull is a dangerous task, for the man must be sure that he can trust his bull and that his training in tying has been complete. There are those who favour breaking the bull first with a donkey. They yoke the two animals together and it is the donkey who leads the bull, for once an ass makes up its mind, it shall not be swayed from its course. In this way the bull learns that he must follow and not lead.

Eric was a good bull and his training moved quickly. I remember in his final weeks we lead him out to the back yard to show Mam. Da stood erect and proud in his white coat and hat, and Eric looked every bit the prize-winner. He learned quickly, it seems to me now, for with little work he followed our simple commands and the cattle stick did not have to be used. In competition, the overuse of the stick is frowned upon, for it shows an ill-trained animal.

On the day of the show, Eric got a makeover. His tail was clipped and his hooves cleaned. He was shampooed and washed, using Da's secret method to make the hair shine, then dried with a special dryer, combed and straightened. Finally, he was given two pints of Guinness to calm his nerves. And so, golden and strong, smelling of rose petal and hibiscus, we took Eric to the show.

I think perhaps my brother and I had the day off from school to accompany Da. The national Simmental bull show attracted men and buyers from as far as England and Scotland, and we were but the newcomers. We were excited by the spectacle of it all. Da walked Eric around the show ring and I remember the feeling of pride hearing the words, 'That's the Connells of Longford.' We did not win any great prize that day, but Eric was sold to a good man for a fair price and Da was noted as a young man to watch out for on the scene. He was not much older then than I am now. In the evening, we ate a burger from the truck outside with salty chips, and went home happy in our day.

I should like to relive those show days again. The purebred has real promise, and we are at a point of decision, for he may well be the best animal we ever breed here, and to sell him could be a fatal move. I have checked that the soft velvet show ropes are still in the tool shed. They are cobwebbed and worn but they could be made great again. We still have our white coats and plastic cattle sticks, too. It could be done again. We could hear men say our name once more at the ringside.

This is Da's decision to make. I shall clean the ropes in hope, like a squire might do his old master's armour.

Crocodile Dundee

There are times in farming when nothing can be done. The beast is too old, the calf too sick, the man too worn out, the situation too hopeless. It is at these times that heads must be cool and hearts must be stout.

The lamb was too big, the ewe too small, the vet was not there and in the end there was but one answer: I had to cut his head off. It is strange to say, but I did not shy away. I knew the ewe would die if I did not act: she would rupture and have an aneurism and bleed out and we would lose two animals. I had seen it once before, with our very first birthing three years ago. It was that birth which knocked Da's confidence and pushed me to take charge. Tonight, when I decided to take up the knife, I asked for the lamb's forgiveness before I began.

There was a small knife but it was not fit to the task and, searching for a time, I found an old butcher's blade and began to sharpen it. All the while Da watching me in a quiet awe. I had skinned animals before. I was the man who took the pelt of a kangaroo and salted it years before.

The blade was sharp and clear and I sawed through bone and flesh and sinew and after a time the head came loose in my hands.

'Them Canadians would never believe you could do this,' Da said, in reference to my old life in North America.

'Aye,' I answered, and pushed the thought of that world from my head.

I fished out the rest of the lamb's body from his mother.

There was space to move now and soon I found his legs and pulled him from her. The ewe shouted and cried but her pain was over and she would live.

'I'll get her a lamb tomorrow,' Da promised.

'That'd be good.'

Sometimes we must take life in order to save it.

Truce

The nights have begun to show on me. I am tired, my bones are sore and my mood is erratic. I wake still weary and I need to rest more, but I will not give in. I do not know why that is – perhaps it is something about needing to prove myself, or my fear of beds, for I have spent too long sick in them.

It is Tuesday and we are cleaning out the cattle houses again. I do not want to do this job today. I feel we are stuck in an infinite loop, a monotony of cleaning and bedding, of feeding and caring of the sick and the dying, and I have not left the farm in weeks and money is tight. We are halfway through cleaning the big house when I snap at him.

'Don't throw the dung into the box. You're getting shite all over me.'

'What's wrong with you today, at all?' Da asks.

'Nothing. I'm fine.'

'Well, you're not in a right mood, that's for sure.'

'I'm fucking tired, all right.'

'Well, take a break, go on in.'

'No.'

'Well, don't be sulking so.'

'I'm not sulking.'

We returned to work then, but before long I shouted at him again for reversing the tractor too close to the wall and we were at loggerheads once more.

'Fuck this,' I say.

'Clean the shed.'

We are shouting now and our voices fill the space.

'Fuck you and the shed,' I say. 'I'm the one helping you when I've work to be at.'

'Ah, you wouldn't know how to hold down a real job,' he says.

'Arra, fuck you, you old bastard,' I say. I never used to call him old, but in our fights I do now. I know this is a low blow when he is still in his fifties, but our tempers are out of control now and the fight will take its course. The animals are looking at us and wondering at what strange sounds emerge from these usually quiet men.

I walk out of the shed, go inside and take a cup of coffee and try to read the paper. I have let myself down; I should not fight with him. I know it is the tiredness and nights speaking. I know too that we are not unique, that fights on farms are as old as the trade itself.

Da finishes the houses himself and after lunch I apologise for my behaviour and thank him for finishing the work.

'I didn't mean that. I'm sorry.'

'That's OK, we all get like that,' he says.

'Think I need a night off from the sheep.'

'Take it. I'll look at them.'

I rest for an hour or so, I watch an episode of *Heartbeat*, an old period drama set in Yorkshire. Later Da tells me there are days that he too wants to throw in the towel.

'But you can't give up when there's over a million euro on the line,' he says.

'No,' I agree.

We end the day as friends. All is forgiven, the houses are clean and the cattle are fine. They have not died without me being there for a few hours. This fast reconciliation is a triumph for us. Perhaps it is a sign of how things can be in the future, perhaps we can find an understanding between us. I hope so. That evening we go in peace.

MARCH

Raft

I smell the hay today and think of summer. I long for its warmth in this cold weather. When I was a boy, the summers seemed longer, the weather brighter. I know now this is nostalgia, that the time was just as short, but the world was a bigger place then, for I was smaller.

Every summer we would build a raft and set sail in the Camlin River, which is the main waterway through the county. Our neighbours, the Lees, would accompany us. We would take a loading pallet and affix four plastic barrels to its underside. Then we'd wheel it down the lane atop my brother's skateboard and carry it across the road and down through Mickey Doherty's fields to the river bend.

We were all of nine and eleven and the work took time. When the raft was floated into the river, we took turns driving it. We would jump from it, fish from the riverbank for trout, and in the evenings fry up sausages.

Mother often came to sit by the bank on hot evenings and soak her feet. She joyed to see us so free and young. We are all of us grown men now; the Lees are emigrants and live in Australia and the North. The children do not build rafts around here any more and Doherty, who was a grandfather to us all, is dead. All things change, even the weather, which we blame now on climate change. But the river flows still and the trout now swim uncaught. I return there from time to time to sit and think. It is a beautiful place. One day maybe the next generation shall sail from there again down the rapids to Kilnacarrow Bridge and steal plums from Trappe's trees.

Inspection

There is a big clean on this morning, for the food-quality people are coming. It is a new scheme from Bord Bia, the Irish Food Board, to ensure the traceability of all meat in the Irish market. Mam has pointed out that the yard needs tidying, and so I have begun to clear the scrap steel and metal that my brother has left here from various building jobs. I neatly stack and pile it upon wooden pallets and move it out of the yard. When the yard is clear, I put fresh disinfectant in the footbath. It has been fifteen years since the last outbreak of foot and mouth occurred, but I still remember the disinfectant at school and the dipping of our feet. The country treated the disease most seriously, for it was highly contagious and could wipe out entire herds. We had seen the damage it had wrought in the UK. We had seen the pyres of burning cattle corpses and did not want this to happen to our animals. In the end, there were nearly 300,000 cows slaughtered in Britain then, there was but one case in Ireland. We are still careful to this day.

Da is inside, preparing files and medicine bottles, for we

must be up to date on the paperwork. The buzzword now is 'agribusiness', for each farm has become a producing unit and we the farmers, the custodians of the land, are now manufacturers, or growers. We have become a cog in the wheel of industry.

Da is not so fussed about the inspection. It will not affect the grants we receive, but, as he says himself, it would be nice to get the stamp of approval, for it might soon be mandatory.

Farming in Europe is subsidized by the EU under the Common Agricultural Policy, which emerged in the 1950s, following a period of food shortage during the war years. Under the policy, EU farmers receive subsidies to farm and produce food for the greater population. Over time, this scheme has evolved to include trade control and Europe-wide compliance to standards such as animal welfare and environmental practice.

It has always been a source of contention amongst the urban and more industrial nations of Europe that farming is subsidized, but, as any farmer will tell you, there is not much money to be made in this profession. The subsidies take the form of the Single Farm Payment now. It is a strictly controlled system and without it many smaller men would be finished.

There are some who argue that these subsidies prevent agriculture in Europe from modernizing, but I do not believe that most Europeans want American-style or corporate farming. People want to know that their food was grown by a farming family in a particular place. I know that the Irish would not opt for big business to run their farms, for the idea of owning our land runs deep in our culture, perhaps as a result of colonialism or the Great Famine, and I think our French and Spanish farming brothers and sisters feel the same way. Our food is better in our own hands.

At midday, the inspector arrives. He is tall and thin and has the air of a true bureaucrat. Da tries to *plámás* him with coffee and some teacake I bought in the local shop. I know better than to be out in the yard and so make myself scare. I will travel to Clonfin and inspect the cows. On my way out, I give Da the thumbs up, but he does not see.

The cows are growing short of grass on the hill farm. We shall have to bring them a bale of silage, but those too are growing scarce. I have not yet counted them, but I think there are only perhaps forty odd left. I do the sums in my head and realize that we will need to buy more feed or let the cows out. But the grass is not yet growing, so I don't know what they would eat.

I do not delay in Clonfin, for the weather has turned, but, driving home, I decide to visit Mary and get the paper. Her shop is the oldest in the village and she and I have got to know each other in the last few years. She is a wonderful woman. I buy the paper or some bread from her from time to time and we talk. She is a great reader and I have lent her books and stories. We talk sometimes of Patrick Kavanagh or Shakespeare.

'It's tough going, but he's a great man with them words.'

'The Shakespeare is beautiful,' I agree.

We discuss the weather and politics and, running out of talk now, I roll up my newspaper and bid her farewell and say

I shall call again soon. I apologise for the dirt my wellingtons leave upon her floor.

'Don't worry, it'll give me something to do,' she says.

The inspection is finished when I return. Da shakes his head and calls the man a pencil-necked cunt. We have not passed, failing on a small technicality.

'We'll reapply,' Mam says as we break for our coffee.

'I guess so.'

'What would he know about farming anyway?' I say.

'Not much. We'll pass next time.'

I unfurl the paper. The scientists have proven Einstein's theory of relativity. Gravitational waves are real, it says. I read it out.

'Haven't I always said that's what attracted me to your father?' Mam laughs. 'He always had a great pull about him.'

I put on the potatoes for lunch and we soon forget the inspector and his visit.

The Wild West

To tell the history of the Americas is to tell a story of bovine expansion. Settlers may have made the Wild West and the frontier, but they followed in the wake of their bovine brother. No other animal has so shaped a culture. So many American icons are associated with the cow: the cowboy, the western, the rodeo, the hamburger, the steak house, the

Marlboro Man, and the very notion of the frontier itself. The story begins six centuries ago.

On Columbus's second voyage, in 1493, he brought with him a species that for ever changed the Americas. Up until this point in history, both the north and south of this vast continent had never met or encountered such an animal. For the native North Americans, this must have been a strange experience, for the cow looked somewhat like the buffalo of the great plains but smaller and near hairless. For the South Americans, it must have seemed an alien.

It is the cow, second only to man, that has colonized all the habitable continents of this world. It is the cow that Europeans brought to their settlements first, it is the cow to which they devoted their greatest shares of land, it is the cow for which they cleared whole forests and species.

In the sixteenth century, the Spanish conquistadors arrived and began to establish ranches in the southern states of what is now the USA, from California to Texas. Their cattle had come from the Iberian fighting stock and it developed into the Florida Cracker (now one of America's rarest bloodlines) and the Texas Longhorn, which recent genetics have revealed to be a true hybrid mix of European and Indian cow species. The Longhorn is a tough and hardy breed. For much of the early colonization of the southern states, the cattle fended for themselves, and so the species evolved to suit its environment, becoming strong, aggressive, fleshy and drought-resistant. The breed was so successful that by the turn of the next century there were several hundred thousand of them throughout the southern states.

Although the cowboy is such an American hero, the first cowboys in the States were actually Spanish *vaqueros.* It was they, along with the Native Americans, who first tended America's cattle herds.

In the early seventeenth century, the English settlers came to America, bringing with them some of the smaller, quieter English breeds. By 1633, the herd of the Plymouth Colony in present-day Massachusetts numbered some 1,500 cattle. For these Puritan settlers, the cow became a symbol of their new life, as well as a link with their agrarian roots across the Atlantic.

Although the cow adapted itself to the American landscape, it can be said that the American landscape also adapted itself to the cow, for it was the cow that opened up the Great Plains. It is perhaps hard to imagine that time now, that frontier, the feeling of the ever-expanding horizon and limitless possibilities of the new country. The Plains have often been called the American Serengeti, and they must have been as amazing to the cows as to the settlers, for they were wide and open and lush.

Following the American Civil War of 1861–5, returning Texans found a state herd of some 5 million. Demand for beef was growing in the north and so the Texas cowboys began to transport the cattle on vast drives through the rich fertile grasslands and on towards the waiting Yankee states. Teams of cowboys would move 3,000 or 4,000 head of cattle at a time. The work was hard, dangerous and brutal. It is the stuff from which legends were born, and place names such as Shawnee, Ogallala, and Dodge City have entered the popular imagination. They were all of them cattle towns where cowboys met railheads. They were places of infamy, where men spent their wages and indulged in the comforts of liquor and flesh.

The story of a powerful steer called Old Blue has also gained a place in the halls of myth. Old Blue was the dominant bull of rancher Charles Goodnight, a bull so good at leading cattle and preventing stampedes that he was used on several trips back and forth across the plains. In his working life, Old Blue led over 10,000 head of cattle. After his last trip, he was not slaughtered but retired to Goodnight's farm, where he lived another twenty years with a small harem of cows by his side, like the Apis bulls of old.

Although there was no spiritual element in the Americans' relationship with the cow, it was an animal that conferred power. Like the Fulani tribesmen in Africa, the early Texans measured their wealth in terms of head of cattle. And like Queen Medb and her Celts, battles were fought over those same animals.

The ox was an important part of this story of the west too, for it was oxen who pulled the great wagon trains across the frontier and allowed for the greatest inward migration of a people the world has ever known. The ox, a castrated bull, ploughed the fields and cleared the forests. It was stronger and more reliable than the horse and mule, it ran on grass, and its dung could fertilize the new American homesteads.

Before long, trainlines were built into the frontier to transport animals. Now there was no need for the great cattle drives. Open-range farming ended and the great enclosures of land began, bringing with it new breeds of cattle that were more suited to fencing and homesteading. By 1885, there were some 45 million cattle in the US; the once plentiful Texan Longhorn had been reduced to just a handful of animals.

The coming of the cattle trains also signalled the end of the nomadic way of life for the native Plains people. As cattle towns sprang up along the route of the railroad, there was an inevitable clash over access to grass for cows and buffalo, and the cow traders ensured that the buffalo lost. The holocaust of this animal, for there is no other word for it, meant the disappearance of a major food source for the Native

Buffalo golgotha

Americans of the Plains. And so, starving and divided, the great tribes were driven off the frontier. Chief Joseph was among them.

To look at the picture on the previous page is to see the industrialized slaughter of an animal on a scale heretofore never seen. In the sixteenth century, there were estimated to be some 30 million buffalo in North America; by the turn of the nineteenth century, there were just 100. But I think this image points at something deeper, something more malign: the destructive nature at our very core. To look at this picture again is to see horrible echoes of human genocide through the centuries. Could we replace those animal skulls with ones of another kind, a sentient kind? Could they be the bones of Rwandans, or Aboriginals, or Jews, the skulls of starving Irishmen or Congolese rubber slaves? The picture I think most of all says to us that what we are prepared to do to one animal can and might carry across the boundary of species, as it later did in industrial proportions in the nineteenth and twentieth century. It says to us that perhaps we practised the art of Genocide on animals first; the buffalo just happened to be that unlucky creature.

If there is one small of piece of redemption to be found in this image of the extermination of the buffalo, it is that the skulls were later ground down for fertilizer and scattered across the soil of the Plains to encourage growth. So they returned at least in some form to what they once knew.

This Lady's Not for Turning

The silage was wrong. I knew I had read or heard this somewhere: we shouldn't have been feeding it to the sheep. This morning one of the ewes has come down with an illness. Though I have not seen it before, I know straight away that it is turning sickness, for she can no longer stand upright nor walk; instead she keeps her head bowed low and moves in circles. There is a general depressed air about her.

Da is away and I am alone. I separate her from the flock and bring her to a private pen I have hastily made. I stand her on her feet, but she quickly falls over. I stand her again and she falls over again and I see now that she will die from this illness. Perhaps yesterday there would have been a chance to save her, but not now, now the bacteria has spread to her brain.

Listeriosis is caused by a bacterium that lives in bad silage, and if a herd comes into contact with a bad bale, you can expect between 2 and 10 per cent of the animals to be affected. I read about the disease on my phone, and halfway I pause, for I can pinpoint the exact bale that has caused this, and I was the one who gave it to the sheep. It came from the front pile and, on cutting off the plastic that surrounds it, I had seen a small amount of blue-white fungal growth on the outer layer. I removed that part and fed the rest to the sheep. But the bacterium must have been alive in it and it jumped from grass to beast and has riddled her. I shall be very lucky if she is the only one.

I am a member of a sheep farmers' group on Facebook

and I turn to my fellows for help, but there are no concrete answers.

I pet the ewe's head and retrace our steps of yesterday; together she and I go back twenty-four hours. She had been walking oddly, I remember, turning into walls, but I had thought her getting ready to go into labour, for she is heavily pregnant and her time is near, and the actions were not dissimilar. I may lose her lambs now, too. Her eyes seem clouded and absent and she may well be blind, for that is also a symptom.

Our front-line defence medicine against all bacteria is penicillin, so I give her a dose. I will call Da so he can get in contact with the vet and see what can be done. But I must be nurse and doctor for now.

Da makes the necessary call, but the prognosis is not good. I sit with her for a time and rub her muzzle. She is foaming from the mouth now – it is sticky and warm. I have put feed and water down for her, but I know she is beyond eating. It is bad enough that I didn't properly look into the question of feeding the sheep silage, and worse that it was I who fed her the poisoned bale. I curse myself for having missed the earlier signs. I clean the foam from her mouth and she begins to spasm. It won't be long now.

I have taken all remaining traces of the feed from the other sheep and it will be hay from now on, even if we have to buy more in.

The thought crosses my mind to shoot her and put her out of her misery, but I have no rifle to hand and a shotgun is too messy. The ewe's breathing is slow and laboured, and then comes the slow inevitable death rattle and the final twitching of muscle and nerve. She died at 5 p.m. Her last hours were not peaceful: blind and shaking in convulsions of delirium she left this world. The bacteria has won and in its victory killed itself. I did all I could, but it was all done too late.

Hunt

The ewe's death, the thought of the gun – it has all reminded me of my time with the buffalo hunters.

It was years ago now, and I was in the bushland of Australia's Northern Territory. The bush of Arnhem Land is vast. Covering some 34,000 square kilometres, it is about the size of France, yet holds a population of only 16,000, most of whom are Aboriginal people belonging to the Yolngu Nation.

I arrived in the far-eastern peninsula town of Nhulunbuy and it was there I met the hunt operator who drove me and the paying customer, a South African hunter called Len, to our camp some five hours inland. There were five of us – six if you counted the cook, but he did not hunt and so was not counted.

The Asiatic water buffalo were introduced to Australia by the British some two hundred years before, for meat and as work animals, but the herds had broken loose and turned feral and there were thousands of wild buffalo now. They were big and strong and looked like the cows of home except for their horns, which were large and black and covered their heads.

Len spoke with a hoarse accent and told me he had paid $10,000 to kill a bull buffalo. The money would be divided between the local Aboriginal community, who gave permission to use their land, and the hunt organizer, a white Australian.

On the second day, we found the herd, which numbered in the hundreds. The buffalo has no natural enemy here save a dingo or crocodile. The Aboriginals sometimes kill them for food, but transporting the animal is so hard that they always have to butcher it where it dies.

Len wanted a bull's head and pelt, and it was my job to record his mission for Australian radio. Our camp was beside a billabong which had a large crocodile. The cook said it was a pet and would never harm us, and he fed it bits of offal. In the mornings, I often heard a mighty clap ring out as it snapped its mouth upon the water's surface. It was over six metres long and far from being some tame dog.

Each day, we trekked through the bush. It was a slow process, for the jungles there are wild and virgin, untouched since creation. Our guide was a New Zealander, who walked barefoot the whole journey. He had not left the jungle for over a year. He told me he would be taking a holiday to America in a few weeks to shoot a cougar. He only went to other countries to kill things.

At night, we ate by the camp fire and drank whiskey. On the third day, the Aboriginal elder came to meet us, for he had heard I was a journalist. His sons carried him from the truck, for he was a leper and had lost his legs to the disease. We shook hands, and I clasped his stump. It would have been an insult not to embrace him and his leprosy was dry. He told me of the buffalo and the coming of the British, which was his word for the white men. He wished us a good hunt and instructed that I come and see his home afterwards.

On the fourth day, we tracked a large bull. The South African told me that he would kill something today. His desire for blood seemed to be a way of trying to exert dominance on the world around him. It felt as if some ancient drive had possessed him, something of the hunter and gatherer, and yet there was something of the consumption of modernity about it too. He wanted one of everything: a dead water buffalo, a dead impala, a dead lion; on and on the list went. Their skins littering his house as reminders of kills past.

The bull was old and we followed him for an hour or more before the opportunity arose to take the shot. We did not speak, and then the crack rang out through the air and the bull was down. He fell as he was crossing a small river. The bullet went straight through his shoulder, smashing both his front legs and heart in the process. But buffalo are big animals and, though his heart was destroyed, he lived for a time. I still remember the blood-soaked breathing as he laboured to deliver those last few gasps. His chest rising and falling, he did not try to move. It was the same painful movement I saw in the ewe all these years later: the agony of death. There was nothing beautiful in it, only a slowing, a fading. When he passed out of this world, we congratulated Len.

'He's a big old bull,' he said.

'How do you feel?' I asked.

'I feel good. I came here to shoot a bull and he's a big old skinny bull, a real warrior, exactly what I wanted.'

They butchered the animal there and then. They cut off its head first and then the guide began to skin it. Its pelt was an inch or more thick in parts. I thought of our cows at home as I watched the bull become a carcass in minutes. We took no meat, leaving it instead for the dingos and wild pigs, who would eat its entirety in a few days, the guide said.

As we returned to camp it was dark and we were tired. We had walked for many hours and our load was heavy. We had to cross a large crocodile-infested lake before making our final turn for home. I was afraid then we would be attacked. The boat journey across the lake took place in silence and I could tell our guide had not meant to come this way. The smell of our pelt and the butchered head would have been enough to drive the lurking creatures into a feeding frenzy. As we approached the shore the motor cut out and my fear surged. Shining our lamps out across the water, hundreds of eyes reflected back at me, waiting patiently for some small mistake. We had the rifles with us, but we knew that the guns were quite useless in the darkness. Above us the stars shone as we rowed the last metres, careful not to touch the giant beasts. When we reached the shore, I was never so glad of earth.

We got drunk that night on Jack Daniels. The bull's pelt was salted and stretched and would be shipped to Len in a few weeks. Its head was boiled and bleached to remove the flesh. We ate sea-turtle meat that the Yolngu elder had gifted us for supper. It was blood-red and tasted like lamb.

At three in the morning we were still drinking and Len invited me to come and shoot a lion for free on his game farm in South Africa.

'Why me?' I asked.

'You remind me of my son,' he said.

'Is he a hunter, too?' I asked.

'He's dead.'

I do not remember the journey home to Sydney. I felt strange. I had not washed in several days. The men had presented me with a hunting knife on leaving. On the plane back to the city, a well-dressed elderly woman sat beside me.

'You look like you've just come out of the wilderness,' she said.

I caught a look at myself in the bathroom mirror later. The bull's blood was still in my hair. I was dressed in a blue shirt and brown khaki, as though in a weird homage to David Attenborough, though I felt very far from being him in that moment.

Organics

I have not been to see the cattle in Clonfin for a week or more, for the jeep is in the garage. Dad has gone a few times in the car, which I'm not insured to drive, and assures me they are OK. I miss the drive up there, for it gives me time to think and reflect.

The listeria was a one-off case, thankfully, and the orf,

too – the sores that had afflicted the sheep and lambs have all but healed. It is hard to say if they are thriving, but they are getting bigger and the lambs are eating their pellets. It has been a hard year. We have been tempted to bring them inside once more, but there's a risk that the orf might return, or another disease. Besides, the shed is full of new lambs now and there is not the space. They are better out. Some days I take them hay and they enjoy it.

The grass is still scarce and no growth has occurred, but we must be patient. We are not alone in this dilemma: many men, many families are caught in this predicament. The foxes are the only things thriving and we have shot another.

The calves are growing, but the weather has stunted them too. Perhaps Da is right that the bull has not bred well. He and Rory have located a new bull and have visited him twice. He is a young bull, Da says, and he has asked the man to hold him.

There are twenty or more calves now. I am busy bedding and watching them. We are constantly on the lookout for illness. I do not want to have a repeat of the ewe or of Red. I had a dream of him the other night.

I was driving in the countryside through a barren landscape, treeless and rocky like the coast of County Clare, where the famous Cliffs of Moher reside. I was pulling a trailer behind me and when next I looked back, it had somehow opened and a cow and calf had fallen out. I was moving at speed and it took time to stop. I did not know how the animals had escaped. When I found them, the calf was drowning in a shallow pool of brackish water – there were small pond-like plants growing in it. I raced to pull him out. His eyes were closed and he was unconscious and so I performed resuscitation on him.

He woke after a time and water fell from his mouth and nose. He was wet, wet as a newborn, and his eyes were black and bright. It was Red and I had managed to save him, in my dreams at least he is still alive. I have not told Da or Mam about this.

Despite the weather, things are good on the farm. Everything is alive and at times such as this I feel sure that I should be a farmer. Perhaps I spent so many years away from this place so that I could finally come back and see it for what it is: my calling. Da and Mam will not farm for ever, so perhaps I will take over. I would make some changes, sure. I would become an organic farmer, for that is the future, or my version of the future at least. If I must provide food for an urban world, then let me provide the best.

As I walk through the shed I dream of what I will do, which cows will take precedence, what breeds. Perhaps I shall have a small Dexter herd for my enjoyment: they are a miniature cow and cute and would be nice in the front field. I should like a herd of Black Whiteheads, as they are good milkers and calve well. It will take work to become organic, but I am prepared for that.

I tell Vivian, my girlfriend, of this dream over Facebook and she laughs. What about your writing, she asks, the reason you are there?

And what about my writing? I do not know. A cow does

not calve in five minutes, and a book is not born overnight. For now, the animals need me and I need them.

Saw

It is time to trim the Limousin cow's horn. She has but one. It has become a yearly ritual now, for it grows quickly and, if left uncut, will grow into her eye. She is an old beast and growing thin at the hips, but she bears a great calf every year and we will not sell her.

We walk her down to the crush and load her into place. I prepare a halter with the rope, first looping it around the back of her head, then under and over her nose. The tying of a halter is an important skill for a farmer to master, for it makes handling the cattle easier. Animals are not handled as much as they used to be years ago and have grown somewhat wilder. What they do in America, God only knows, for the cattle are never really housed and have become part-feral.

I tighten the rope and loop it around the front metal post of the crush; when she is secure and in place, I nod to Da, who takes up the saw and begins to cut the tip of the horn. He is good with a saw, and with wood too, from his years in the building trade. I've never had his handling of such tools. In the summer, he designed and built a long hall for the playschool. The timbers were two-hundred-year-old pine from an old Orange lodge in Northern Ireland, which he planed and sanded and cut. The hall was beautiful when it was done.

'I could never do this,' I told him.

'Ah, sure, it's not that hard,' he replied.

The school children love it, too. He was proud that week, happy in using his old skills.

The saw is nimble and quick and Da is careful not to get dust or bone flakes into the cow's eye. Their horns are made of bone, but closer to their head there are veins in it.

The tip of the cow's horn is gone now and I unloose her.

'That'll keep you for another while, girl,' Da says.

I lead her back to the shed. Her udder waddles as she walks, her time is nearly up and she will calve soon.

Neighbours

Farming is a hard life and our fellow farmers understand the journey. We can share with them our losses and triumphs in a way perhaps we would not with others. We are a community here and a good farmer is lost without his neighbours, for, though we perhaps would not always admit it, we need one another. No man, no farm, is an island. At times, our neighbours will be stuck just as we are: a broken-down tractor, a man short on a cattle move, in need of an extra trailer, or just a friendly ear and advice.

I know other men's fields and ground as well as our own,

for I have helped and worked them too when they have been in need.

In the summer, I helped stack the square bales at Murphy's, for their sons are emigrants now and they had not the help. It cost me but an hour out of my day and they were glad to see me come round the corner. I remember too when Uncle Mick died and the neighbours came to make the silage. I do not know who instigated this, but men came with tractors from all over, the meadows were rowed and baled and the feed wrapped and stacked. I have never forgotten this act, nor the men who did it.

The importance of neighbours came to us a few weeks ago in all its glory, for the McVays, whose land *mearn*, or abuts, ours in Kilnacarrow, had a horse stuck in a bog hole in the moorland that meets both our properties.

Try as Jim might, he could not release the mare on his own, and his tractor had not the power to pull the animal from the embrace of the swampy earth.

It took our tractor and four men to free the animal. She cried and neighed as we worked to free her. The fight had gone from her body and she had not the strength left to help us in any way. Our slings and ropes missed and failed several times and at the last, when all our hopes were faded, Da and I tried the sling one last time and we all pulled her free.

We were all of us united in the joy of our actions for that short moment. She would have died there that day if it were not for Jim and us, his neighbours, and for that we are all glad.

Canada

It snowed the other night when I was out looking at the animals: I saw it fall from the eternal blackness of the dark. It is strange that it is snowing now, but this winter is not yet over. It fell in patches, melting as it hit the wet ground. I stood for a moment and watched it mysteriously drift. It brought Canada back to me.

The winters were long there, the snow high and heavy. By the lake house where we sometimes stayed, the waters would freeze over and we could drive snow mobiles across them. I remember too the sight of a bear in French River, and a golden eagle by the waterside. Nature was vast and epic there. I dream of it all sometimes still.

My old love and I have not talked in a year or more, and I don't know what could be said now. Things feel so different from how they did this time last year. I was back on the farm last winter too, but I found no joy in it then. I did not count myself a farmer that year, just a labourer looking no more than a day ahead, a wanderer trying to find his way home. That winter was the darkest I had ever known, though the weather was not of nature.

I look into the night and put out my tongue and taste the snow. It is cold and fresh. A cow bellows low and clear from the shed and rouses me from my thoughts. It is three or four in the morning; even the bats are asleep, and I must do the same. I don't remember the time difference with Toronto any more.

Property

The Celtic Tiger boom brought with it money and dreams. Some saw us farmers as backward then, for working our land was far less profitable than selling it. We could have sold sites for development, we had fields with road frontage and they would have made us money, but the money would be gone now and the fields would be full of new houses.

There are still plenty of empty ghost-estates in our county, half-finished suburbs which stand as memorials to another time, evoking some strange future that never was. The recession did not just leave these empty buildings upon our landscape; it forced thousands to emigrate and drove some, in the face of financial ruin, to suicide.

Mam has told me that Da sometimes thinks about the bank repayments for Ruske's and has suggested that we sell Clonfin to pay the debt sooner. Now that the recession is ending, there is an appetite for building once more and selling the hill farm would clear the bill, but the land that has been with us for so long would be gone. We did not want this to happen during the boom and we do not want it to happen now.

I have not spoken with Da about this Clonfin idea. He knows that one day I hope to build a house there. At times, I have imagined my sons and I working the fields, as he and I work them now. My vision of the future has changed many times, but I still think perhaps it will come to pass.

Charters

'Charters has his lambs out,' I say.

'How do you know?' Da asks.

'Saw them when I was out on me bike.'

'How do they look?'

'They're good,' I admit.

'The hoor never has bad ones,' Da says.

The Charters family have lived in the area for over two hundred years. They have been here as long as we have, but we are of different tribes. I only came to know old Charters in the last few years, for I set about to write a story about the assassination of his uncle Willie during the War of Independence.

The war with the British will never be forgotten in Ireland. Running from 1919 to 1921, it was both the defining of our new nation and the end of the old one. An Ireland died in that time: British Ireland. This was the Ireland in which the Charters family lived as members of the Protestant and Anglo-Irish community.

Willie Charters was but a boy when the IRA killed him, dumping his body into Lake Gurteen. Old Charters cried when he told me the story. There was nearly eighty years' distance from the event in the telling and still the emotion was real. Willie had been innocent, he said; he had been no informer. And Old Charters began to explain how his uncle had got caught up in the conflict.

Following a land dispute between two members of the

Charters family (brothers William and Robert had argued over how a parcel of land was to be divided between them), they turned to the newly formed IRA courts for justice and a decision in the matter.

The IRA courts were the only functioning law of the land during the war years. They met in a local hall and made their proclamation. The court's judge, a Mr Victory, decided that the fields would need to be divided properly, and so set about to walk the land with other members of the party and the Charters family to conduct a fair survey.

But at the arrival of the IRA men, the young Willie, the son of William, left his family and informed the British forces of the enemies located at their house. Victory and his aide, Barney Kilbride, were arrested by the British as enemy combatants.

The IRA commander of the area, Seán Mac Eoin, treated Willie's actions as the most serious act of treachery. It was he who ordered his capture and execution.

Willie Charters was younger than I am now when the IRA men blindfolded him and put the pistol to his head. Da says he was given a choice to be shot or drowned. He chose the lake, but they shot him anyway for good measure. The killing changed and shocked the Anglo-Irish community in the area. There is a statue to our war leader Mac Eoin in the village now. We do not talk of Willie.

I have finished my short story on Willie but have not given it to old Charters yet. I must make sure to do so, for he is getting on now and it would, I think, gladden his heart to know that Willie too has not been forgotten.

I tried to give him something in fiction that reality did not: dignity in his end.

Clonfin

Mam has come home from Clonfin and she's furious. The cows are roaring and thin and there is no grass. Her anger is directed at me.

'I've not been up there for a week or more,' I say. 'Da said they were fine.'

'They're a shame to me,' she says. 'What will the neighbours say? They need a bale, they need nuts.'

'I'll take care of it. I'll take care of it.'

Da is gone to the mart and it is just as well, for the row would have escalated.

'I've no one to take care of this farm. I should just sell it,' she continues, shouting now. 'Your brother, you, your father, none of you is a farmer. I'd be better without it. You can't even take care of the animals we have. I'm going to sell the whole thing up, I'm fed up of this.'

'I'll go up there now,' I say.

Her words have hurt my pride, for I have been working day and night on this place. But I am also angry with Da, for why did Da say the cows were fine? Why did he not tell me the cows needed grass? I have done nothing wrong but I must be in the middle of all of this once more. I spit and shout and Vinny jumps on my leg to be patted, but I shout at him too to sit down. He puts his tail between his legs and runs away.

'Fuck it, fuck it.'

To feed them, I will need to bring a round feeder with me, a type of metal holder for feeding animals outside. Its metal exterior prevents the animals from tramping their silage into the ground.

First I load its large split metal halves onto the rear of the tractor, for I will assemble it when I get to Clonfin, then I put the blade on the front loader and take a large wrapped silage bale and now I am ready.

I think of Ma again and sigh. These are not her true feelings, I tell myself. She is upset and this will pass, even if it takes several days. Perhaps it is this weather, but the fights have grown more frequent lately, the unease in the house has come back. The work is wearing on all of us – the nights, the calves, the sheep, Red's death. We are at each other's throats. They tell me this is the way of farming, but I do not like it. When I am in control, I think, I shall do things differently. I shall not fight with my sons. And in the same breath, I curse the whole thing and think I'd be better to return to the city and not bother with any of this shite.

It is raining as I move along the country roads to Clonfin. The tractor is slow and the drive takes forty minutes. I have time to cool down, to cool off, as does Mam. Why can things not be easy, I ask myself.

The cows are waiting for me at the gate when I appear. They low and bellow but they are fine, they look wet and rained upon but are not starved or perished. I know now that the strain is showing on mam too. That the weather has affected her too.

I drive up the long slow hill and unchain my feeder and assemble it. The ground is wet and the muck comes up to the top of my wellies. I trudge and labour and a light mist falls from the sky, but soon everything is where it should be and I can load the bale. I unfurl the plastic and load the silage into the feeder and the cattle stand around it now and they gorge themselves happily.

I listen to an interview with a famous American writer, Jay McInerney, as I work. As he speaks, I see the towers of New York, the world of East Egg and the jazz age of Fitzgerald. The interviewer asks him about his youthful success and he laughs, and I am thoroughly relaxed now and have forgotten my woes.

I do not say anything to Da when I come home. There has been no crisis and I will ensure the cows get a bale every few days to keep everyone happy. While there is silage still, I shall feed them. Soon the spring will come and we will move them off this scrubby ground at Clonfin. The meadow grounds of Ruske's and Esker will be lush and green and we will all be able to breathe easier. Today has shown me one thing: that all of us need a break.

Hitler's auroch

It is a strange thing that a beast that had been extinct for over 8,000 years could occupy the mind of the Nazis, but then they were not ordinary men.

Hitler's fascination with genetics has been well noted. His Nazi ideology was underpinned by eugenics and the idea that the impure and the unclean would be removed from the genetic pool, that effectively selective breeding would create a pure Aryan race.

These fantasies extended, however, not just to people, but to animals too. And so it was that the auroch, the majestic beast, was to be exhumed from the past and resurrected by Nazi scientists. Its story is a sad episode in the history of the cow and one we should not forget.

In the 1930s, the politician and military leader Hermann Göring recruited a zoologist named Lutz Heck and his brother Heinz to bring the auroch back from the dead. This ancient forebear was to be bred to be hunted by leading party members, so that they could recreate the old German legends, wherein Siegfried, the central hero of German mythology, hunted the aurochs, amongst other primeval European animals, including elk and wisent (a type of buffalo).

Such was the commitment to this romanticized past that a location for the newly resurrected herd was identified in advance of the conquest of Poland. The aurochs would be homed in the Białowieża Forest, Europe's last extant wilderness, which already held stocks of Eurasian buffalo, elk

Heck cattle on Schiermonnikoog island

and wolves. It would serve as the exclusive hunting ground of the German elite.

When the Germans did finally take the forest in 1941, 20,000 of its residents – many of them Jewish – were rounded up, expunged or deported. The Jews of Białowieża were amongst the very first victims of the Holocaust, cleared from their homes to make room for an Aryan cow. The Jewish community of Białowieża was wiped out.

Working independently, the Heck brothers created their aurochs through a process they called 'breeding back', crossing several breeds of European cow – including Spanish fighting bulls, Highland cattle and primitive breeds from Corsica and Hungary – to draw out their muscularity and strength. The programme was not able to replicate the massive size of the original aurochs, but their characteristics were exactly as the Nazis wanted: aggressive, powerful, dangerous and wild.

Although the aurochs were eventually introduced to the Polish forest, the Allied victory in 1945 ensured that the great Aryan hunts never took place. The defeat of the Germans in 1945 also signalled the end of the cows. The partisans and locals destroyed all the creatures they could find, seeing them as an abhorrence, a symbol of evil, a living Minotaur.

Today, the Heck aurochs number just 2,000 and they shall not be mourned when they die out.

Despite their support of the ethnic cleansing of Białowieża Forest, neither of the Heck brothers was ever charged or found guilty of war crimes.

It is Shabbat today. I don't much want to look at the cows any more.

Luck

Our luck has changed. I cannot put another word to it. The gods that rule farming have been kind to us with the cattle and the sheep, but now, now they have turned their backs on us. The Irish are still a superstitious people, but at times such as these I too believe. I have interceded to Saint Francis, lit candles, but still the deaths come. It has begun its harvest and I am powerless.

It began when I missed my alarm a week ago. For twelve nights straight, I had been on call. Twelve nights where there was only the yard and my incessant walk back and forth through the darkness to check on the sheep. Thankfully the cows were not calving as well, not yet, at least.

My body was worn out and I slept through the 3 a.m. alarm and the 4 a.m. and when I finally woke at six I walked out to the yard to find the best lamb of the year dead. He lay in the open ground, still wet with the fluids of birth, twice the size of any of his fellows that were already weeks old. His mother cried and pawed the ground in an attempt to make him rise, but you cannot raise the dead. The sight of his perfect but lifeless body broke me in a way, and I questioned the very nature of life, the spark, the divinity, the soul, and its absence from this lamb. His loss brought up others in my life and my heart was suddenly sore and I was no longer sure who I was mourning for. Perhaps it was the lack of sleep, perhaps something else, but, like his mother, I now tried to will life into him once more, hoping that, like some Lazarus, he would rise from the straw. And why not, I thought; didn't Christ begin in a barn, a stable?

'Fuck it anyway,' I muttered, for these thoughts were madness.

I carried him from the large pen and placed him inside an empty fertilizer bag. I cursed myself for sleeping through the alarm, for I could see that if I had been here, he would be alive. I think of all the mistakes I have made, of the loves I have lost, the lives I have left behind, Canada, Australia, roles I have inhabited and abandoned, the journalist, the producer, the writer. And now, as farmer too, I have failed. There was so much wrapped up in this lamb. His death was the counterpoint to the calf I had delivered triumphantly by myself at the start of the season. The dark to its light.

'It's not your fault,' Mam said when I returned inside.

'It is my fault.'

'You were tired. It can't be helped,' she said. 'If they're going to be lucky, they'll be lucky.'

I say nothing.

'You've had such a good run with them,' she added.

'I'm going back to bed.'

It was still early and I was exhausted. I slept until ten but I did not dream.

When I came out again, Da didn't say anything about the missed alarms and there was no fight. He too knows the emotions of death and understands the loss. The sheep, these cows, they're something we share.

'It's only a lamb,' he said.

But we know it is so much more.

Death the Leveller

The week has not improved. There have been four more deaths. First, there were twin lambs I delivered, but they were small and premature and the birth was difficult. Their mother was a first-timer and this brings with it problems. It took me more than thirty minutes to take the first and when he came out his neck was broken. I do not know if I killed him in the process. The second lived for a pace, and even suckled, but he too passed a few hours later.

When the next set of twins died the following day, my heart hardened. They had come in a jumble of legs. And I thought for a time I should have to cut them out.

Soon, the fertilizer bag was full of their lifeless forms. I took the bodies to the knackery yard. I have seen too much of this place this year.

On Thursday, we had a day without death and we thought our troubles ended. But at noon on Friday, it began again.

One of the younger ewes went into labour and she too was expecting twins; she was a good strong Suffolk and I expected her to carry. But after twenty minutes, her water sac failed to emerge. In the birthing of a lamb there are different stages, and if any should fail, we know that something is wrong. We must be vigilant, with Suffolks especially, for their lambs are big and strong and so harder to birth.

This ewe had ring womb, which means her muscles had not relaxed and her cervix would not open. With careful massage and liberal use of birthing gel most ewes will dilate and a normal assisted delivery can occur. Rubbing and spraying and massaging, I began the procedure. I moved quickly, for I knew time was against me. I would perform this chore for five minutes, give her a rest for five minutes, and then repeat. But after the manoeuvre was finished, she was still tight and clenched, I could fit but two fingers inside her and I knew that she would never now open.

'There'll be no lamb come out this way,' I said to Da.

'What do you mean?' he replied.

'I mean she needs a vet. There's nothing I can do, she's locked down tight as a jail.'

Broken now too from our week of failure, he fought with me.

'She'll open herself.'

'I'm telling you she won't. You need the vet or you'll have two dead lambs.'

After more cursing from me, the vet was called. I was scared now that the ewe was also in danger and that in this death would claim a triptych.

The vet was a stranger to me, a woman from the North, from another practice, for Gormley was not available. She instructed that the ewe be placed on her side and her feet bound. I did as I was told and then began to pluck the wool from her flank so that her skin was visible and clear. It was pink and raw like that of a man. A small pool of wool littered the pen. I rubbed the ewe's muzzle and with that the vet began her work.

'She's a fine ewe,' she said.

'Aye, she's not too bad,' Da answered.

I did not speak, for I did not want to jeopardize the operation. After the last few days, I had invested everything in this moment, in the hope that life might be rescued and our failing fortunes reversed.

The scalpel cut the various layers of skin and then we saw the first lamb. Slowly she took him from the open womb and handed him to Da. He cleared the mucus from the creature's nose and then shook him three times to clear the fluid in his lungs. The lamb shook his head and cried and we smiled, for he was alive. Da handed me the lamb and I poured water in his ears to fully wake him to the reality of this world.

The vet returned to her work and pushed and cleared and took the second lamb and again the procedure was repeated, but our smiles turned quiet as the first and then the second lamb began to fade. I performed chest and heart massage but both died.

'Ack, Tom, I'm sorry,' the vet said, and she began to close and sew the ewe shut.

We didn't speak much after that. I walked the vet back to her jeep. Da put the lambs in the bag. We counted only the cost of her visit now. We counted only the mistakes we had made. We did not fight any more that day, for the dead were not worth that.

'If the ewe makes it through the night, I'll get her a lamb,' he said.

'Right,' I replied.

We waited for Sunday. Next week would be better we told ourselves. It was all we could say.

Overnight

The bad week is over at last and I am staying with Granny tonight. Uncle Davy and his family have gone down to their caravan in Sligo for a night away and have asked me to stay at the house with her. I am not minding Granny, for even though she is ninety she needs no help.

'It's just the company I like,' she says and she prepares a supper of fried eggs for us.

'That's no bother, Granny,' I say.

'These are my eggs,' she tells me.

'They're lovely.'

After we have eaten and supped our tea, we sit down to a good hour or two of solid criticism of the world. Granny is getting tired of the 1916 Rebellion celebrations. It is a hundred years since the beginning of the Irish journey to freedom and though Granny is the last woman to receive the widows' pension for the war, she says the whole thing is overkill.

'We'll be sick of it at the end of the year,' she laughs.

'We will.'

The general election happened a few days ago and it seems the county has no Member of Parliament, which annoys her, too. She made sure to vote, but she will tell none of us who she voted for. That is her business. When we have run out of news, we begin to discuss the neighbours and talk of who has died.

Granny listens to the death notices on the local radio each day. It is part of her life, for many of her friends have gone now. We never talk seriously of her own mortality. I joke that she shall live to be a hundred; she says she hopes that is not the case. I do not like to think of that day, but I know that she is quite ready for death.

After the news, we watch a traditional music programme on TG4, the Gaelic language station. The Chieftains are in concert and we enjoy the jigs and reels. We smile and point out the players we know. We allow 'O'Sullivan's March' and the 'Foggy Dew' to roll over us. It is at times like these that we know who we are: this music, the language, is part of our legacy as a people. Granny's people were all musical; her brother was an all-Ireland winner on the mouth organ. I have never asked if she can play or sing, but I should not be surprised.

We talk of old times before bed. She knows that I am in a way collecting the memories of the long ago for the next generation of the family. We discuss again the story of Kate Mullen and the *Titanic*. Granny has the telling of it and, like Da, she is a natural.

'Did you ever know her, Granny?'

'We got her letters and parcels home, but she never came back herself,' she says.

Kate Mullen was but a young girl when it happened. She was on her way to America to make a new life in New York and work in service. It was 1912 and Granny was not yet born.

When the ship went down, Kate was trapped behind the metal gates and it was a local man called James Farrell who freed her. He sacrificed himself in the process. Hollywood scenes have been made around that moment. Kate was placed in the last lifeboat just before the ship sank and in his final act James Farrell threw her his hat, saying, 'I shall see you in eternity.' Like so many others who perished on that cold dark night, James's body was never found. Kate lived out her days in Long Island and never set foot on the ocean again. She passed in 1970, an old woman.

'She dined out on that story for many a year,' Granny says.

In James Cameron's film, the character of Tommy Ryan,

the Irish friend of Leonardo DiCaprio's Jack Dawson, is based on Farrell. His family came from America when they unveiled a monument to Kate Mullen in our local parish, right by Father Seán's house. They were good people. They had not forgotten their roots.

'They called it the unsinkable, but it went down,' Granny concludes and switches off the telly.

'We'll get an early night.'

'We will.'

In the morning, she has already fed her hens and has breakfast waiting for me. We have tea and salute the new day. I linger for a while and we talk of our sleep and what work I have planned. As I leave I joke that she is the unsinkable lady of Soran. Long may she sail.

A Break in the Clouds

We have brought home some of the cows from Clonfin, for the rain has gotten worse and they are not happy. They were not hard to move and were happy to return inside. We will feed them extra nuts until they are ready to calve.

The flooding has subsided in the west of the country and the daffodils have begun to bloom along the lane. Spring, it seems, is here at last.

The first batch of calves are thriving now. The illness that has dogged them has lifted, there are no more cases of scour, and for that we are thankful. They have become bold in their age now and play and prance in the yard. They chase and are chased by Vinny. The cows must smell the change in the weather, for they too seem restless. The fields are still wet, however, so we must be careful when we release them. Soon they will all be out in the meadows, but not yet.

Our fodder is holding. I have counted the bales and I think we have enough to last for another month, as long as we are not wasteful. We also have hay enough for the sheep.

The second batch of lambs is coming on well. We decide to move them over the road to the small paddock in front of my brother's house. It is fenced for sheep and they will not be able to escape. Not all the land is fenced for them, and I remember our first attempt at sheep all too well. It was years ago now and I was but a boy. I was our dog then too and chased those same sheep through fields and ditches. It was a disaster.

'We'll put them out where we're set up,' I tell Da.

He nods and agrees.

I take feeders over for them and I will continue to take them nuts to aid in their growth. Each bag of feed costs 10 euro, but it is money well spent, for I can see the change. The lambs in the upper ground are growing strongly now, too; they are thirty kilos or more. They eat their feeder empty every few days.

We have faced what this winter has had to throw at us and though we are damaged, we are not broken. I allow myself a brief smile. The work has been hard and the rain long, but perhaps things are at last on the up.

It will be a few days until the next cows start calving or the third batch of lambs begin to arrive. I can allow myself to rest. For a week I have no night shift to attend. The bags around my eyes begin to fade and I look more human again.

Mam is playing her rebel songs on the radio and I know too that she is in a good mood. Everywhere there are little miracles, graces of spring. Blooms are appearing on the early flowers and trees, like messages from heaven, signs from the other world that life will continue for another year.

I come in from the yard and take a standing mug of tea.

'There's a kindness in that day,' I say.

'We could be over the worst.' She smiles. 'What have you on for now?'

I list my chores and suggest then that I may go to see the Heslins. It's been a few weeks since I saw them last, before the height of the calving and lambing started.

'They will have wondered what happened to you.'

I agree and leave her to the rousing chorus of 'A Nation Once Again'.

The Heslins

Willie and Deirdre Heslin are dear friends. They are the parents of my best friend, Liam. We have known each other for nearly twenty years now. I've been calling over most weeks since I have been home and it breaks both our schedules.

They too are farmers and we share our trials and tribulations. Liam has chosen a different path in life and is an actor. It was a big step for someone from rural Ireland to strut the stage, but he has been driven to do so since we were boys. He is on a world tour at the moment. I would be jealous if I were not so happy and proud. His rise has been as if it were my own, for we both dreamed of the time things would fall into place and our big break would come.

'That's a grand day,' I say as I step out of the jeep.

'Arra, John, we thought you were dead,' says Deirdre.

'I told her it was the bloody sheep,' says Willie, smiling now.

'It was. And calves, too,' I say.

'Don't talk to me – we're sick tired of the same things.'

We move into the kitchen and Deirdre puts on the kettle for coffee. We will sit and talk and share our news. The Heslins live in the next parish and so support a different football team and shop in a different store. To hear their news is to hear from another people.

'How's the calving going?' I ask now, as we sup our drinks.

'Willie had a nice calf born last night,' says Deirdre, as she lights a cigarette.

'A bull,' says Willie.

'Great stuff. Good calf?'

'Our bull is breeding well. How's your stock?'

'Good now, but not as good as other years. One loss.'

'You'll have that,' says Willie firmly.

'Had a lad touch and go with pneumonia, but he's coming to again,' says Deirdre.

I ask after Liam and tell Deirdre what news I know of him. He is busier now and we all don't speak with him as much as before. He is doing well, though, and Deirdre is happy. Sharon, her daughter, is in Limerick finishing a master's and so their house is quiet.

We talk of politics then, in which Deirdre is well versed, but she can offer no solid facts on which way the negotiations will go to form a new government for there has been no clear winner. The country is without a government and it shall remain so for some time.

Biscuits and sandwiches are produced and we turn our attention to sport. The club football is in a resting period with only minor matches occurring. We talk of the rugby and the demise of the new Irish team. It is not the side it once was and it has saddened us. Willie keeps me abreast of the soccer, of which I know little.

Soon an hour has passed and our talk has run dry. Willie walks me down to the shed to show me his calves. His weanlings are just about to go to the mart. I compliment him on them. Liam will not be a farmer, he has told me, so I do not know what Willie will do when the time comes. Right now, he has life in him yet and he knows no other work. He is proud of Liam and sees that he is making a living at something he loves.

'How are the books going?'

'The cows are more profitable,' I say.

'It'll all fall into place.'

'I hope so.'

I wave goodbye to them and promise to call again more

regularly now that the bulk of the calves is born. We joke and laugh as I pull out of the driveway. They have been great friends to me in worse times. I have not forgotten that.

A Bovine Revolution

While the Americans were exploring their Plains and allowing nature to produce the Longhorn and Cracker breeds, an agricultural revolution was occurring in Britain that would change the world for ever.

It began with the enclosures of the Tudor period, which put an end to the medieval system of open field farming, which had seen large tracts of common land divided up for strip grazing and cropping. This system was feudal in its nature and did not make full use of the land. The enclosures had

Meadow Landscape with Cattle *by Willem Roelofs 1st*, c.*1880*

many knock-on effects, most notably the fact that landless labourers were forced from the commons, thereby creating a newly mobile work force who would eventually make their way to England's northern cities to fuel the Industrial Revolution.

Private ownership, or the enclosing of the fields, also allowed for large farms to develop, and with that came improved farming methods, for with their title deed in hand farmers could innovate freely. The process of enclosing the land faced initial opposition, but as agricultural output began to grow faster than the British population disquiet became muted.

The British agricultural revolution was not solely caused by the enclosures, rather a combination of factors came into play: the development of the crop rotation system, which allowed for more agricultural output than with the old system of leaving fields fallow; the creation of new ploughs and other farming tools; the establishment of a national market to sell the output (it was here that the self-regulating market concept took root); and finally the success of selective breeding.

It is to one man that the cow owes the next step in its history. Robert Bakewell may not be so well remembered today, but without him there would be no Charles Darwin and no Alfred Russel Wallace. Born in 1724 to Leicestershire tenant farmers, Bakewell travelled extensively through Europe in his youth and studied various farming methods before returning home to work alongside his father until the latter's death in 1760. It is at this moment that Bakewell the scientist was born, for he began to experiment with his whole farm, dividing and irrigating soil and fields, and pioneering the process of selective breeding.

Up until this point, different farm breeds had traditionally been kept in close geographical proximity and breeding occurred in local areas. This method of farming had produced regional variations, but certain characteristics belonging to a great cow or a superior breeding bull might be lost in the next generation. Bakewell began to separate the sexes, keeping male from female, and controlled the breeding process himself.

This method of 'in-and-in' breeding was a form of artificial selection: Bakewell bred bulls to cows he desired and at times inbred bulls and daughters to preserve and thus promote certain traits. Bakewell was in effect wresting control from mother nature and selectively breeding his cattle and sheep.

It was Bakewell who is credited with producing the first beef cow. The Dishley Longhorn was the result of crossing English Longhorn heifers with Westmoreland bulls. The resulting breed was able to meet the increased need for beef in the industrializing cities of Britain allowing in part for the workforce to grow in numbers yet again, to some 32 million by the beginning of the nineteenth century – a jump of over 26.5 million in two centuries.

Bakewell's 'in-and-in' breeding programme was so influential that his breeds and ideas were transported across the settled world, from Australia to North and South America. With this new programme, the size of cattle was transformed too, and cattle weights on average doubled, to 380 kilos, within two decades.

Selective breeding or artificial selection was to play a

profound role in the culmination of Charles Darwin's theory of natural selection. Writing some sixty years after Bakewell's death, Darwin cited the agricultural pioneer's work directly, stating that the artificial selection showed variation under domestication. In other words, Bakewell demonstrated evolution in action, by identifying and breeding certain traits and characteristics, thereby leading to the creation of new subspecies. Bakewell had done in his lifetime what nature had taken millennia to achieve. It is thanks to Bakewell and his cows, as much as to Darwin or Russel Wallace, that we owe the origins of our now accepted views on evolution.

So profound was Bakewell's influence and ideas that new cattle breeds began to appear around the world, including the Charolais in France and the Shorthorn in north-east England. Interestingly, a century later, the British breeds of Hereford and Aberdeen Angus were bred smaller, using selectivity to suit the market demands – so small in fact that a full-grown Aberdeen Angus reached only to a man's belt buckle. The breeds have since been bred up again and now stand at nearly twice their nineteenth-century size.

Sadly, not all of Bakewell's new breeds of cattle and sheep have remained popular, including the Dishley Longhorn, which is a rarity today. This has perhaps more to do with changing fashion and taste in cow breeding as with beef production. Like the great fashion houses of Paris, even farmers are victims of trends.

The cows of my childhood, the cows of our farm, we owe to Bakewell, that quiet champion of change. It is easy to view our cattle as wild animals domesticated, but in reality they have been carefully bred and nurtured to shape our needs.

Spin

The day of the big charity cycle race has arrived. I have been training for several weeks now and feel strong and ready. I will do the seventy-kilometre event; it is a longer route but the challenge will test me. Da has said he will drop me and my bike in the local town at the start line. Cyclists from all over the midlands have come – farmers, tradesmen, professionals – and we are all here together to race and salute the man who died founding it. I soon find fellows that I know and we talk and laugh and discuss the race.

We take our positions and I band together with an old school friend I have not seen in years. The race begins and we pedal off. We will travel across the countryside and into the next county of Leitrim. The weather is fine and bright, there is a briskness to the air but we do not mind, for soon we shall will be hot and it will cool us.

'How are the cattle, John?' my fellow enquires.

'They're alive and kicking. And your own?'

'Grand now. It's been a long winter,' he says.

'It has at that.'

We cycle harder now and move from the motorway to country roads; we talk of sport and women and history. It is a good chat.

'We haven't seen you down the pub in a long while,' he says.

'No, I stopped drinking,' I say.

'Good man, I've often thought of doing that myself. Has it been hard?'

'Not so hard when you put your mind to it,' I say.

'And how are you after everything last year? You're feeling well?' he enquires.

'I'm in good health again and glad of it.'

'You can say that again. I thought it was good that you were open and spoke about it.'

'It hits many families,' I say, and, happy now that I am OK, he leaves the subject to one side.

At Cloone, part of the race group breaks off to take tea in the village hall. By the football pitch the men have arranged the letters of the scoreboard to wish us luck. People line certain corners and wave and we return their smiles. I have brought biscuits and share them with those around me to keep our energy up. And then we're off again. As we reach Mohill town in Leitrim my fellow tires, for he has not trained so much and so I slow and keep pace with him.

'You can go on,' he pants. 'I don't want to hold you back.'

'We started together, we'll finish together,' I say.

'That'd be good.'

Leitrim is the home of the small farm. It is a small county and the land is poor, but through that pressure they produce some of the country's best cattle. It boasts five livestock marts for a population of but 25,000. It is positioned perfectly between the Republic and Northern Ireland, and so traders from Belfast and Derry come here to buy cattle.

We turn now for the rugby club and the finish line. Our talk has become less, for our bodies are worn. We look forward to the food that will be waiting for us, we talk of Terry, the man who is not here, the man who started the day.

'He'd have enjoyed this crowd,' someone says.

'He'd be in the leaders' pack,' another answers.

'The good are taken too young.'

We lament and mourn. We do it in the way the old ones have shown us; though we may have iPhones and listen to dance music, we are still countrymen with country rituals. In our every pedal now we remember the dead, in our every pedal we celebrate life. I am glad to be alive. I am glad to be well.

The club house is full of racers when we return. We have two and three helpings of curry and then I cycle home, waving goodbye and farewell.

'We'll meet at the next race,' I tell them.

'We will,' they reply.

Black Cowboy

I remember meeting my first cowboy. In Australia, farms are called stations and cowboys are known as jackaroos or drovers, but the settlements are just as vast, the terrain as unforgiving and the weather as harsh and strong as in America.

Old Jack was a retired drover, an old Aboriginal man who had spent his youth working on cattle stations throughout northern Australia. He was a member of the Gurindji tribe, a nomadic group of desert people whose traditional lands comprised over 3,000 square kilometres in the Northern Territory. In the 1850s, the arrival of British settlers and cattle men brought with it increased competition for water, as well as brutal massacres. Faced with such odds, the Gurindji took to settled life working on the white-owned cattle stations.

Jack had flown in from his home to the regional town centre of Katherine to attend some meetings. He was dressed in a cowboy outfit of a check shirt and jeans and he wore a battered Akubra hat. His accent was clipped and broken, for English was not his first tongue. We sat in one of the local pubs and talked of cattle and life and farming.

The harshness of Australia's interior scared off most white settlers and so many of the stations hired Aboriginal stockmen and domestic help. They were cheap labour, working for little money and minimal food rations, and living in tin shacks, known as humpies, with no running water or sanitation.

'Life was hard,' Jack told me.

His face, as I can see it now in my memory, was lined and worn, but his smile was one of kindness that I had not seen in his white contemporaries. I think this smile perhaps came from his past, for he had taken part in the Wave Hill walk-off in 1966, during which two hundred Gurindji stockmen, house servants and their families went on strike. This was the start of the land-rights movement in the Northern Territory and the beginning of Aboriginal people standing up for their own human rights.

Wave Hill station had belonged to the Vestey family. Baron Vestey was a British cattle and shipping magnate who owned large tracts of the Australian outback, and who had been responsible for driving the Gurindji and other tribes off their traditional land. Not for the first time, the cow had been used as a means of dispossession.

After a beer or two, Jack began to tell me of the walk-off. Their leader had been Vincent Lingiari, a stockman who led the workers to a sacred site nearby at Wattie Creek, where they began their seven-year strike. What had started as a movement for wage equality turned into a demand for the return of Aboriginal land from the Vestey company. The company and later the Northern Territory Government attempted to buy off the Gurindji with bribes of cattle meat and wages, and later employed bully-boy and intimidation tactics, but eventually public support grew too strong and the Australian Prime Minister, Gough Whitlam, himself negotiated with the Vesteys to return part of the homelands. In 1975, a handback took place and the Australian land-rights movement gained its first major win.

There are not many black cowboys any more. Jack and his kind are the old fellas – they listen to country music, drive four-by-fours and swig a beer on a hot day, but the great cattle droves of their youth are over. The farms of the territory are run by helicopter and motorbike now, and for many of the free generation of Aboriginal people welfare cheques have replaced the need for work, and with that has come social problems.

Jack is a rarity now and his life offers a glimpse into an older world. He told me he was still a pretty good judge of a cow, and we discussed the benefits of the Hereford and

Brahman cross. He told me of the need for drought-resistant cows and laughed when I told him that we must bring our stock inside during the winter.

'It never gets that cold here,' he said.

We shared a steak, which is cheap and plentiful there, and bid each other farewell.

That was a long time ago now. I do not know if that black cowboy is still alive. He has stayed on in my memory, like the red dust of that place and the dry, cool nights. Once or twice I have thought of him on hot days here on the farm, when the earth has hardened and cracked and, for those brief few days of Irish summer when it does not rain, we think that we too are in the desert.

Sting

There has been another calf born and all was well, so we have quietly agreed that things have turned the corner.

In the week of the bad luck, I had nightmares again: this time it was not Red the calf, but aborted lambs. They were deformed and not fully grown, they cried out blindly and were covered in red fluid. I woke in fright that night. The image of those lambs has stayed in my mind, as has the loss of their real brothers.

Da has come down with an illness. He was stung by a wasp several times on the leg and the wounds have turned sore and red and he has been in bad pain with them. After the second or third day, it was agreed that he needed the doctor, and so Mam took him. It was a queen, the doctor said, for that was the only wasp that could be alive at this time of year. How she ended up in his trousers, he does not know, but the queen's sting can be bad. The doctor has put him on antibiotics to prevent an infection and given him some cream to bring down the swelling.

He does not look well and I have said that I will cover things until he is back to speed.

'It's not manic out here, I'll cope,' I tell him.

'If you're sure,' he says.

'I'm sure.'

I do not like to see him laid so low, and it makes me think of the future, when old age will keep him out of the yard.

'Cunt of a wasp,' he says.

'A right bitch.'

'There's a few westerns yet you haven't seen of them films I got. Stick one on and throw yourself on the bed,' I say.

'Aye, aye, I might do that,' he says.

I return to the yard and continue my jobs. I do not mind working alone and I have become used to it this winter. Despite it all, despite our fractious relationship, he has taught me everything I know, everything that I need to be a farmer.

Lost

I'm a lamb short. There were fourteen lambs on Monday and now there are only thirteen. I have walked the small paddock in front of my brother's house four or five times. I have searched the ditches and looked in the surrounding fields, but I am a lamb short. I am still working the farm on my own, so I have no one to consult with, but he has been missing two days now and, with no sign of a body, I have to face the fact that he has been killed by a fox.

I call my brother, who is a good hunter, and ask that he keep a watch with his rifle over the lambs. He agrees that he will. He does not ask about the missing lamb and I do not say. I do not like to admit my loss, not when I am running the farm on my own.

The others are faring well. I bring them nuts every day and then walk to the back of the hill, where the purebred calf is located. He is growing fine and strong and his sight lifts my spirits. His foster mother and brother greet me and I feed them too.

Whichever ewe's lamb went missing, she does not seem to have noticed, for they are all lying out now on the hill, basking in the early spring sunlight.

I shall have to watch them closer. I shall have to outsmart the fox.

Cows on Canvas

I watched a film in the sitting room this evening. It is our good room and we seldom use it, save at Christmas or Easter. Above the mantelpiece sits a painting of a rural scene of cattle drinking by a river stream. I do not know the painter, but pictures such as these hang in so many country houses. Constable is a favourite. Nowhere has he found a greater audience than with the farmers of England and Ireland, for in his works do we see ourselves. In our townland (the small land divisions of a parish) of Soran alone, there are more than four reproductions of *The Hay Wain*.

Since the time the aurochs were first depicted in the cave paintings of France and Spain, it seems we have sought to capture the magnificence and strength of the cow in art. To look at this gallery of cows is to see the different meanings we have placed upon them through the centuries of understanding.

Of the depictions in Egypt and India I have mentioned the sanctity of the cow but not perhaps the aesthetic beauty of the works. These stone carvings and stone sculptures tell us something about the eye of the creators and the intent of their makers.

In European art, the ox most frequently appears by the manger in nativity scenes, but it also features in other Biblical depictions. Poussin's *The Adoration of the Golden Calf* shows the cow on a plinth being celebrated by the Israelites, and in the painting by Rubens and Brueghel, *The Garden of Eden with the Fall of Man*, a cow nudges into the right hand side of the scene.

From the Renaissance onwards, the ox and cow are depicted as working animals, a symbol of the northern European Protestant ethic; later, in the work of Van Gogh, it seems they are weighed down by some other burden, perhaps guilt or sheer exhaustion.

In the twentieth century, it is perhaps Picasso who has best captured the power of the animal. To him, the bull became a fixation. I have seen whole gallery wings full of his works from the early to mid 1940s, when he saw the bull as the symbol of power and recklessness – in it was death, the fiesta and the shadow of fascism. Indeed the 1942 found-object piece, *Bull's Head*, is made from a simple bicycle saddle and handlebars, and yet with these man-made items and this man-made work of art, he is also celebrating the natural world, at a time when nature itself was being pushed out of the way.

Guernica, Picasso's masterpiece of the Spanish Civil War, also features a bull. Every day for five years as a secondary-school student I ate under a reproduction of it in our school canteen. I did not know it was a Picasso work then, only that it featured animals I knew; the horse, the bull, even its lamplight, felt familiar to me. It was to me like a chaotic scene from some stable or outhouse. A fight bigger than any thirteen-year-old could imagine.

It seems now, on looking back, that cows have surrounded me in my creative life.

In our hallway there is an old print of Millet's *The Gleaners*. I never knew it was a famous work of art until we saw it on the television years ago. Mam said she bought it as it reminded her of farm work in years gone by.

I wonder now, will anyone ever paint our cows? As a youngster, I tried to capture the calves on paper, but my hand was not quick enough and their patience too short. The painting in the sitting room shall have to do. Those painted herds look peaceful in it.

APRIL

Round and Rounds

My days are full of farming now, from morning until night. Da is still sick and moves from bed to living room, but he is not well enough to come out. I would not mind the days but for the nights, which have begun again.

Our second set of lambs are coming, so there are now midnight deliveries and 3 a.m. feeding times. Newborn lambs are like babies, for they must be minded and cared for, they must be fed and watched. There has been another calf born, too. The cow did it herself and for that I have thanked her, for I had not the power in me that night. He is white with red patches and seems good and strong. When I went to help him suckle, he was not interested, and although I have been watching to see if he has sucked by himself, I have missed it. But he is alive and it has been two days now, and so I surmise that he has been born with the right instincts.

My morning rounds are longer now. I begin at Clonfin and check the remaining cows have enough to eat – they are getting through two bales a week. Then the sheep in the upper ground near Uncle Mick's old house, and the second flock by my brother's. The purebred calf and his family must also be fed, and when all the morning jobs are done it is nearly one in the afternoon and I must begin the lunch for the family.

The bags have returned around my eyes and yet I like the order and purpose. I was not always so disciplined a person, but I have found that it suits me and that in the order I can achieve much.

I have tired of podcasts now and so listen to audiobooks. This week it is Hemingway's turn, and he has already transported me to the mountain peak of Kilimanjaro. In Vinny's barking I have imagined I hear the laughing of the hyenas. Today, I am listening to *The Old Man and the Sea.* I have heard it before, but being so land-locked here, it gives me a vision of the ocean. I fancy that my task of running the farm during this time is somewhat like the old Cuban fisherman's struggle with the great marlin: a long and arduous effort, neither of us knowing what will be thrown against us. It has been many months since I have been by the sea and perhaps when all this is done and Da is back, I will go to the coast.

I clean and bed the cattle houses; I have now lost count of how many times I have done this, for there is no point in keeping records. Shit is eternal: it always keeps coming and it would break a man to recount how much he has had to shift or move. It is better to face each day in the moment.

The cattle we brought home from Clonfin are getting close to calving and more new life will be on the way.

The bales are dwindling now every day. There is just enough left, but I may need to buy in feed. I have discussed this with Mam and she has agreed. We are not alone in this shortage – other farms are suffering from the wet winter too.

It is dark by five and I have spent all my daylight hours

on the farm. I call Tim, my friend, for a chat and retire for a few hours.

Loser

It has been over a week now that I have been alone on the farm and the last few days have been stressful. Another lamb went missing and I suspect a calf has pneumonia. The weather has also turned wet once again and I am constantly tired. Da says he is still sick from the sting and I have not questioned this, for I know that if we stay apart, there will be no fighting.

He has come into the habit of asking me to do certain things from his sick station and I have carried out his requests. Today was to be no different. He felt it was time for the third group of lambs to be let out with their mothers. I agreed and said that I would do it. It started simply but it ended in chaos.

Before releasing the lambs, I injected them for fluke and worm and checked their weights. They were all of them strong and fit. I opened the gates and doors and walked them and their mothers up the lane by the river to the upper ground with the rest of the sheep. I was careful and slow in how I released them, for I had learned my lesson on this and I did not want any estrangement to take place. Some tracked back and forth, but, bit by bit, I hushed and coaxed the new group towards the others and after ten or fifteen minutes they had mingled with the big herd and together they numbered some 130 or more now.

It had been raining for a while and I was already soaked through when I got back to the shed. It was then that Da came out and asked why I had released the sheep.

'You told me to,' I said.

'I meant let them out to the small paddock here by the house. Them lambs are too small.'

'You never told me that,' I said.

'Any bollox would know not to let them out to the upper ground. They're not fit for it yet,' he growled.

'I did what you asked me to do,' I said calmly.

At that point, one of the newly released ewes returned to the shed. What she came for, I do not know, but her lambs were with her. She baaed and Da spat and then he shouted that they would all have to be brought back down and put in a separate field.

And so we began. The rain grew heavier and more intense and, bit by bit, we brought the newly released group to the small paddock. We did not speak then, for I knew he was in a rage and to speak now would only add to his anger.

I was in the lower ground by myself when the phone call came.

'Get up here now, you bollox. There's a sheep lying out up here, how did you miss her, huh? She's nearly dead.'

'What? Where are you? I'll come up with the tractor.'

'I'm up here, in the upper ground,' and then he hung up abruptly, as he does when he is in a temper.

I did not know where exactly he was and so called again,

but he would not answer me and so now my anger also grew. I started the tractor and put the box on the front loader so as to carry this sick sheep. I had walked the fields that morning and had noticed no illness, but perhaps I missed her. Sheep are not so hardy as cows and sickness can take them suddenly. I hoped then it was not turning sickness, for I didn't want to lose another ewe that way.

He met me on the lane and shouted at me to get out of the tractor. Although my anger was strong, it had not taken hold and I let him into the driver's seat and walked down to open the gate for him.

I heard him mutter, 'What sort of an idiot could miss her?'

Suddenly I could take it no longer and months of suppressed rage inside me spewed forth. I was not bowing down this time but standing up for myself. I stood in front of the tractor then and told him to talk with me, but he would not face me. Again I stood in front of the tractor and told him to talk with me and again he would not. He began to drive the machine forward then, but I stood in its way, refusing to move. I was determined that we would talk this day, rather than shout at one another. We would solve these fights once and for all. He made to drive over me then, but still I stood my ground and began to shout to him.

'What in the hell is wrong with you? I've been running this place day and night for nearly two weeks and then you come out and start picking holes in everything. Talk to me, will ya?'

'Running it? It's a mess, sheep in the wrong fields, a sick animal.'

'We were getting on fine the two of us. What the hell has gotten into you?'

'We're not getting on fine, and it's a mess,' he said again.

'That's not true, I'm here to help you, you're getting older, it's too much for one man.'

'No one asked you to help me.'

'Mam did.'

'Well, I'm telling you now, I don't need you.'

The rain was beating down upon the tractor and it was soaking my back and legs. I felt a stab in my heart at these words, at all the work I'd done, and the lives and illness we had faced together. I stepped up inside the tractor then and forced the engine off.

'I've been working here for the last four months and you tell me you don't need me. I've other things to be at, I could be working on my writing.'

'Your writing! You've wrote four books now and none of them have succeeded. You've no job, no money, your life is a mess and you're a failure. You're thirty and you've nothing to show for it. This is all a repeat of last year, and what did we come out with from that? Nothing! Your mother worries about you still, and she doesn't worry about the others. You're going to turn around one day and be sixty and have nothing. You will have wasted your life on those books and this farm. I don't want you blaming me that you couldn't achieve your goals because your time was took up at farming. I don't need you.'

We were silent a while and simply eyeballed one another.

'Then, there's nothing more to say,' I said, quietly now.

'Nothing.'

I should have cried then, had life not made me a stronger man. It crossed my mind to shake his hand to say goodbye, for it all seemed so final, but I did not. Too much had been said to now forgive or see a way to forgiveness.

I walked back to the house in the pouring rain. I had lost my cap and my hair trailed into my face and my glasses fogged over with the beating, pouring winter rains. I looked back as he drove on into the mists, shaking his head.

The Past

W. G. Sebald called it 'the dog days', Winston Churchill called it 'the black dog', and long before either of them, it was known as the black bile. It was only after the fact that I came to name it. I call it simply 'the Past', and it has shaped my life. I was not in love with life then. I did not run or swim or cycle. I did not see the joy in each calf's birth nor kiss fuzzy lambs' heads.

I spent six months in a cold bedroom, unable and afraid to leave, grappling with the very concept of life. Things had been good before this, but of course they always are.

After leaving Australia, I had made a life in Canada with my then-fiancée. We lived in a penthouse apartment and I did not want for money or material things, for she was wealthy and I was starting to make a name for myself as a writer and film director. Before the fall, life was so different.

Albrecht Dürer, Melencolia I

When the Past arrived precisely I cannot say. It came upon me like a weighted cloak of water, pouring over body and soul until I felt that I was drowning.

As a teenager, I had been fascinated with the works of Albrecht Dürer, in particular the print of *Melencolia I.*

It was odd that this etching, created in 1514, should

have held the attention of a sixteen-year-old farmer's son. Why I loved it so, I cannot say, for I didn't even know what the title meant then, but I drew and copied it many times and attempted to create my own screen print of it in my high-school art class. Raised on films about different forms of creative genius, I knew that all great men suffer for their craft, and I associated this figure, and deep and troubled thought, with achieving something worthwhile.

It was said that before he created the image, Dürer confessed, 'What is beautiful I do not know . . .'

The image of that etching returned to me two years ago and haunted my dreams, and I understood then what Dürer meant.

Our relationship was ended, the wedding was cancelled, and the life we had planned came crashing down, and I found myself once again back in rural Ireland.

A great unease and sadness had come upon me and I could no longer see the beauty in anything. I pushed everything away from me in a sort of madness as I rebuked life and love and battled with the elements of darkness.

I stayed in my bedroom, emerging only occasionally to help on the farm when the lambing was very busy or a calf needed to be delivered. The farm then seemed as a prison to me, and one I longed to escape. I neither spoke nor listened to anyone. I still remember the abhorrence of that rural Christmas, alone, afraid, and thinking only of death and of wanting to leave this world. I was a man possessed with a deep and sickly sadness.

I have a friend, Charlie, who is a faith healer, and it was he and modern medicine that saved me. Through a combination of the old and new worlds and ways of healing, the six months of darkness ended and I emerged reborn. I came to know that not until we are lost can we truly find ourselves and, in so doing, we can then measure that thing men call life and living and appreciate its ordinary bounty.

I wrote of my experiences in articles and in the farmers' paper, and ended up on national radio talking of the Past, of my fall, of my journey through depression and mania, and of mortality itself. Why did I choose to speak out? Perhaps because I felt I now could. I supposed it was a postcard, a letter, from the other side, a message to those still suffering to hold on, to say that life does get better.

Days after the broadcast, a man sent in a letter to the radio station saying he had been on his way to take his life when he had heard me speak. He had been driving in his car, with a rope and medicine for an overdose in the backseat, ready to kill himself, and I had popped onto the airwaves. My words had stopped him from that ultimate act. And he had written in simply to thank me for saving his life.

I cried then, for surely, I reasoned, all of it had been for a reason. Had I not walked that path through grief, depression and the struggles of mental health, his life would have ended that day. Such is the interconnectedness of this world.

It was shortly after that I met Vivian again, after several years apart. It was a wholly unexpected thing. But then as the old book says: 'To everything there is a season and a time to every purpose under heaven.'

And now I have returned to this farm, to the place where I

was so tortured and sick, to write, and it is no longer a prison.

On learning the secrets of perspective from the Italian masters, Dürer published them in a book, for all of his northern contemporaries to read and understand. So it was with me. I have talked and written and revealed the secrets of perspective, the perspective of melancholia. It has helped people. It has helped me.

I return now to that question of Dürer's: What is beautiful? It is, I think, life itself and living. In this farm, I have found my Walden, my sustenance. I walk its fields and know I am alive.

My friend Charlie, the healer, says that I'm meant to write, to create some great work, but I wonder now if the great work I had dreamed of as a teenager, copying Dürer's etching, might instead be a simple thing, an elegy to the nature I know.

Scale

Despite practising one of the oldest professions in the world, farmers are often the early adopters of new technology. As farmers embraced new methods of farming and breeding and began to use steam-powered machines rather than beasts of burden to work their land, so agriculture was industrialized. As output increased, so the world's population grew, and so too did the need for more beef.

It is to America that we must look once more in the story

of the cow, for it was here that its modern chapter began.

There are now 1.6 billion cows in the world, one for every fifth person. Through selective breeding, man began to produce the meat or milk he desired, but it was through science that this process was perfected.

The development of vaccines and antibiotics in the first two decades of the twentieth century meant that large groups of animals could now be housed together indoors, without fear of bacterial infection or spread of sickness. As our largest domesticated animal, the containment of the cow in a small area had huge advantages. It cut down on farming time, and, with controlled feeding, animals could be bulked up and brought to kill weight sooner. By the late 1960s, the advancements in machinery allowed for feeding houses to be built, and with that factory farming began.

Farming in America soon became less a family enterprise than a business endeavour and by the year 2000, 80 per cent of America's beef was produced by just four main companies. The Plains and the range had given way to concrete and steel.

Factory farming, or Confined Animal Feeding Operations, works on the basic principle that maximum output should be achieved on the lowest possible cost input. This method has ensured that the world has had enough beef in recent years (albeit often of lower quality), but it has come at a price – that of lower animal welfare. A confined beast is not a happy one, but selective breeding and artificial insemination have allowed industrial farmers to produce more docile and husbandry-friendly animals.

In this the cow has been lucky, for it has not been the subject of such experiments or such drastic physical modification as its fellow, the humble chicken. Theirs is a life of hardship: housed, caged, de-beaked and slaughtered. If the reality of factory farming has a mascot, the scrawny, mutilated chicken is it.

Recently at a farming conference, I listened to data scientists and academics talk about the development of intensive farming in the future. Agriculture, they said, was the last frontier of the tech sector, and it was an exciting time to be in the industry.

In years to come, it seems, quality won't be a concern any more, because we will all produce the same type of animal. Consumers demand cheap meat and it is our job to make it, to feed the supermarket-driven race to the bottom. Machines will weigh and dispense feed, low-paid workers will carry out whatever manual work must be done. The people closest to the animals will not be farmers but line workers.

The factory farming of cows has not yet taken hold in Ireland. The cows here are still largely pasture-fed and housed only in winter, when there is no grass available naturally. To the tech entrepreneurs who see agribusiness as the future, I and my fellow farmer are Luddites, relics of the past. And yet, I am happy to continue in the old ways, as are most European farmers. Under EU farming law, the use of growth hormones in the production process has been banned and the use of antibiotics must also be tracked and traced; in this way we prevent the flow of animal antibiotics into the human diet and food chain.

Growth hormones are not the only problems of industrial

farming, in the 1980s and 1990s the emergence of bovine spongiform encephalopathy (BSE) threw light upon the problem of bone-meal feeding. In this practice, which many European farmers used as part of intensive feeding programmes, cattle were given bone-meal feed, which contained cow and sheep leftovers from the slaughtering process, as a source of protein. Traditionally, US farmers used soybean meal, but the bean did not grow well in Europe and so they turned to bone meal. The cow, of course, is a herbivore and in the eating of its own kind, nature fought back with the development of BSE, more commonly known as mad cow disease.

The disease affected the animal's brain and spinal cord, eventually rendering it unable to walk and driving it insane. The suffering of the cow was just a foreshadow of the danger of intensive farming, for when this illness crossed the species barrier from animal to man, it took the form of variant Creutzfeldt-Jakob disease, or vCJD. This disease caused memory loss, dementia, depression, hallucinations and eventually death. At the height of the epidemic in the 1980s, it was discovered over 400,000 infected cattle had entered the human food chain. Millions of animals were slaughtered to eradicate the disease, but sadly some humans had already been infected with vCJD and, to date, there have been 177 deaths in the UK. The UK's beef market, which was the worst infected, had exports suspended for ten years and although the ban is now over, British beef remains tainted in some markets.

The long-term risks of intensive farming are not yet

known, for it is still in its first generation, but the BSE epidemic was a warning that caution is necessary. Perhaps in future our industrial meat will come, like cigarettes, with a warning: FACTORY FARMED: EAT AT YOUR OWN RISK.

The industrialization of the cow itself is not just about meat anymore but the valuable by-products of its entire body. Indeed the real value of a cow is in what's known as fifth quarter; the offal, hide and organs of the beast. Fifth quarter is where the processors make their real money. The entrails and offal are harvested and used to produce over 200 products from insulin to face creams; in modern agribusiness no part of the animal is wasted. The meat factories do not pay farmers for fifth quarter and in so doing deprive them of most of the real value of the animal.

Meanwhile, feedlots have grown in number and there are now 82,000 in the US alone, supplying most of the meat to American consumers. Even fortress Europe, with its traditional farming, is under threat from imports from Brazil, the South American beef giant. Farming there is a high-intensity business: huge tracts of rainforest have been cleared for the non-native animal and in the doing, soils and environments have been ruined, indigenous residents driven from their homelands, and vast areas of verdant woodland turned to desert.

Globally, the rise in cattle numbers and increased amounts of methane gas in the air are contributing to climate change and a rise in the planet's temperatures. Greater numbers of cattle also require a greater use, or misuse, of water. It takes around 15,400 litres of water to make a one-kilo piece of meat, and in some countries that water is far too precious. It seems that the Minotaur has awoken from its millennia-long sleep and escaped its maze, razing the environment as it goes.

Through its relationship with man, the cow has been transformed over the millennia from a mythical animal, a carrier of gods, a maker of galaxies, to a carefully programmed 'product' in the food chain. Where once the cow was man's most valued companion from the natural world, now its value in some nations depends on removing it entirely from this world. It has lost its sentience, it seems, in the minds of those involved in the industrial process.

The End of the Line

We have not spoken in a week. The next crop of calves and lambs are coming. He asked me to come out to deliver a lamb, but I refused, for he would not apologise and so we are at an impasse. Mother feels the strain the worst, but understands that I must stand my ground and that I cannot just ignore what has been said. She understands, but she worries too.

'What's going to happen to the farm?' she says.

'I don't know, I don't care,' I say.

'You do care,' she says.

'What's the point? I'm not wanted here, all my work was for nothing.'

'There'll be no one to take over this farm. I'll sell it. I'll sell it all.'

I feel guilty at hearing this, but it was not me who abandoned the place.

'If he said sorry, I would help him.'

'Your father never says sorry, and he never admits he was wrong.'

'Then I can't help him.'

'I've no one to carry on this. What was it all for?' she laments now.

She leaves the house and returns to work.

The lambing is not going well. Mother has had to take over the night shift.

Intercession

It has now been ten days and I have taken other work. I did not tell my friend of the fight, only that I needed work. In the evenings now, I spend my time in my room and eat separately from him. He has lost several lambs. He is still unwilling to ask for my help and I shall not give it, for I have been too deeply hurt.

On Tuesday evening, I met him with the knife in his hand, he had to cut the head off a lamb stuck in labour. Partly I lament for him and partly I think it serves him right; let him realize what I have been doing on the farm.

The bales are running very low and I can hear the cows are

roaring with hunger most days now. A calf is very sick with pneumonia and he has begun to stomach-tube him fluids to keep him alive. He is too ill to suckle. He may well die. I miss them. I miss my animals, and I don't like to think of them suffering because of this human quarrel. The rain is still coming. The nights are wearing on Mother. Someone must give.

'You have a kind heart, John, you're forgiving,' she says.

'There's some things I can't forgive. Even our Lord lost his patience once.'

I think of Jesus then, of the Saint Francis painting hanging in the shed, of the Brigid's cross I have woven, of the litany of divinity around this yard and house, pagan and Christian, and I reflect.

'These are just cattle rows, John, every father and son has them,' she tells me.

I nod for she is right: it is the way of farming and has been for centuries. But I am still sore and hurting.

John McGahern had trouble with his father. It has been well noted. I once heard McGahern refer to him as mercurial and he told how they had come to blows and never spoke again. I know that we too could have come to blows and that everything then would have been ruined beyond repair.

I wonder now, do the cows notice my absence? I do not know. Do cows think of their masters? I cannot say. Perhaps they shall hold a meeting for me, like Orwell's *Animal Farm*, calling out in the animal tongues for my return.

Writing

It happened when I had given up hope. It happened in a dark loft, amongst the insulation and the sweat. My work is to be included in a literary magazine and already a publisher has made enquiries about publishing my book. I wiped the dirt from my hands and face and read the email once more. When the day's work was done and the cash in my hand, I told Mother.

'That's great,' she said. 'It's finally starting to fall into place.'

'It is,' I said.

I did not tell him. How could I?

When the photographer came to take my picture for the literary journal, he insisted I not bring them out to the yard. Those were the first words he had spoken to me in nearly two weeks.

The bales are all but gone now. Mother has asked me to get some more and I agreed. Whatever our differences, the animals should not be made to suffer. I drove to Clonfin one of the days and gave the cattle their food. I did this not for him, nor to ease his workload, but out of habit and to see the animals once more.

The new bull has arrived. He is young and strong, snow white with a plash of darkness around his head. I am not decided on the animal yet. His head is different, he is a Charolais, yes, but he has a different look from the old bulls. He seems too to have a temper. A wicked bull is of no use on a farm, but perhaps he is afraid and getting to know his new home. These things take time.

He took the old bull to the mart. Mam tells me he got a decent price for him.

The launch of the literary magazine is to take place in Galway, and my friend Duncan is coming from London to go with me. It has been many months since we have seen one another. We are to take a trip down the coast afterwards. It will be a break, a break from here. And at last I shall see the ocean again.

'Why don't you invite your father to the launch, John?' Mam asks.

'I don't want him there.'

'Are you embarrassed of him?' she asks.

'No, it's not that. What would we say to one another?'

'I don't know,' she admits.

I am finally starting to be a writer and this is what I have always wanted. And yet, am I a farmer? The sight of the animals now hurts me, for only they know the work I have put into them and the emotions I have been through. The farm, the scene of my victories and defeats, has come to feel like taboo ground. I am not sure if I belong there any more, if I have any stake in its sheds and fields, its sheep and cows.

Duncan

The weather is improving. There are still four bales left and, instead of buying more, he has decided to release all the cattle to the fields. He did not ask for my help. The rain has stopped but the fields are still wet; the cows will find grass but they will destroy the ground in the process. Another week or two inside would do no harm, for we still cannot be sure what way this weather will go.

I have returned to the yard in a way. We do not work side by side, but rather I carry out my own jobs with the sheep, with the cows. I have cleared the scrap metal and rubbish that has littered the place from the winter. The calf with pneumonia was still sick. He had lost parts of his hair and his tongue hung limply from his mouth and I thought he might die. I helped stomach-tube him, but we did not talk during this process. I felt for the animal, for I knew that we would have to stop tube-feeding him soon or we would scar his throat and weaken his sucking muscles. There is only so much medicine can do and then nature must take over. By the fifth day, he drank from his mother. I do all this now not for Da but for Mother, for I know the stress has grown too great for her. It is not fair on her being in the middle of all this.

The day of the launch has arrived and I am suited and cleaned and Ma has lent me her car for the drive to Galway. In the mirror, I look every inch the writer in my suit and glasses, but I do not feel it. I feel this is all pretend. I am only an actor, a farmer on his day off, a farmer without a farm.

The west of Ireland is a beautiful place, the woodland hedges of the midlands give way to stone-wall country and cows are replaced by horses, for this is the homeland of our horses. The soil in the west is poor, but it is here that our culture survived the most. It was here too that Oliver

Cromwell supposedly drove the conquered Irish after the wars of religion and genocide of the mid 1600s. It was he who forced over 50,000 defeated Irish into indentured servitude in the Caribbean. The Irish have remembered this as an act of slavery. There are Afro-Caribbean people in Barbados and Montserrat today with red hair and Irish names. They are our lost tribe, our lost county.

'G'day mate,' Duncan says, as I park the car in the city and we hug each other. We have not seen each other in over two years. I met Duncan in Australia, and he has been a good friend and confidant ever since. He too is a writer, though once he was a trainee doctor. His Australian accent seems stronger now, amongst a sea of Irish voices.

'You all ready for the launch?' he asks.

'Ready as I'll ever be,' I say.

'Did your folks not come down?'

'No, they're busy,' I say, and do not go into the matter.

The night is a success and I am introduced as the farmer-writer. I speak and am spoken to and I wonder if I will spend my life in these literary circles. Patrick Kavanagh, the poet, left his farm so that he could become the great man of letters; Seamus Heaney left his beloved Derry farmhouse; even Henry David Thoreau's Walden ended. I buy Duncan a pint and we decide that we shall leave the city in the morning and head south to the Burren. The weather is with us and the day is promised fine.

'A few days from the farm will do you good, mate,' he says.

I nod and agree.

The Sea

I have so missed the sea. We have driven our small rented car south from Galway city, towards County Clare. The road hugs the coastline and soon we enter the lunar landscape of the Burren. It is grey and rocky and unlike anything I have ever seen. Billy Joel plays on the radio and we sing along.

Duncan is a keen runner and we agree to find a place to stop and lace up our shoes. By the village of Fanore we park. The sun is setting and, though I am tired, we begin. The country roads are quiet and the evening is beautiful. We are in our T-shirts and running tights and look every inch the Yankee tourists.

It is the fourth or fifth kilometre before Duncan begins to talk.

'How are things at home, John?'

'Ah, they're OK . . . There was a big fight,' I confess, and unfurl the details of the row.

'I thought so. You seemed a bit dull.'

'It was a lot to take in.'

'Well, you'll be out of there soon.'

It was then that we passed the cows. The herd peered over a stone wall, looking at us curiously.

'There's a red Limousin,' I say, and point out the beast. 'They're fiery but bring an easy calf. And there's a Belgian Blue, they've got good, lean meat.'

The cows moo and peer at us and I continue to name all the breeds and explain their qualities to Duncan.

'You know, it really is beautiful here,' says Duncan.

'It's not such a bad place at all,' I agree.

'When do you think you'll go back to Australia?' he asks me now.

'You know, I'm not sure. I've come to enjoy the farm.'

'But you surely don't want to spend your time there, not after everything that was said. You'd be much better in a city.'

'Maybe . . . But, now that I'm back, I realize that this is what I know. It's what I've always known.'

We ran on then for a time in silence. We turned a corner and jogged by the seashore; a cow and calf nibbled sweet grass to our left and I smiled. They looked peaceful and at ease.

We ran on into the sunset, tired and happy, the sound of our running shoes clapping on the country road. That night we ate by the Cliffs of Moher and fell asleep to the sounds of the Atlantic Ocean crashing outside.

The King

Duncan and I have seen several small islands on the horizon on our trip. Some are still inhabited and I tell him so. It has brought back memories of the king who came to Soran.

There is a small island called Tory, off the coast of Donegal, and when I was a boy, its king launched a mission to find a mate for a cow on his island. She was the last of her kind, a near-extinct breed of Shorthorn, and with her death would go the end of an ancient bloodline.

After searching for six months, the nearest genetic match for the cow was found to be a young Shorthorn bull called Felix. He was owned by Tommy and Jack Brady, the bachelor brothers known as the Wild Men of Soran. They lived on the brow of Soran Hill, they were farmers and shopkeepers, they were our friends.

When they agreed to sell Felix, the islanders banded together to purchase him. It was the talk of the county, for the television cameras and newspaper men had come from Dublin to record the event. The Tory people saw the re-establishment of the cow line as imperative to their livelihood and tourism trade, it would provide fresh milk for them and allow them to export safe BSE-free beef.

The islanders treated it as a marriage and presented Felix the bull with a laurel halter woven with wild flowers. Tommy said Felix was going home, for the bull's grandfather had been a northerner, the Letterkenny Eagle.

The bull and cow lived happily for a time and the Wild Men gained a new fame in the area as breeders of great bulls.

Duncan laughs at the story. I am glad to have its telling.

MAY–JUNE

Writer and a Farmer

We finished the trip with a mighty run through a forest. Duncan was faster than me and tore off into the distance. Now that I was alone again and our time together was coming to an end, I knew I had to make a decision. He would go to London, but where, where should I go?

Despite my talk of the cows on our run, I cannot go home.

When Duncan left Ireland, I did too, and travelled to Spain to meet Tim.

In my self-imposed exile, I began to write again. I had been all year waiting to do this, but when I finally sat down, it was not the western, nor the literary novel I had planned, but something which has been within me all my life. I wrote about our cows, from the little Black to Red, from the Master to the new bull. And I wrote about Da and Ma, from whom so many of my stories have come.

It was strange to write this, to delve into my memory and look at what has happened. In a way it allowed me to make sense of things.

Father Seán once told me that fiction is a truth that never happened, and truth is a fiction that did. I have long pondered that statement. Now I think that somewhere in the middle, narrative is born and with it meaning. As I wrote, I began to see that life is just a series of events, and that it is we who shape it to explain it to ourselves. The calves had become so much more to me than mere animals; they were part of the cast in this battle of wills, in the age-old story of fathers and sons. I have reflected too on that day in the rain and Da's hurtful statements, and wondered if he spoke not just to me but to himself, for maybe he wished he had made different choices, too. Or maybe that he wanted me to feel free to walk away from the farm, to do something different with my life. Is that not what he worked so hard for, all those years on the building sites, all those late nights, that back-breaking labour? To give me – to give us – a chance at a different way of being. To become a man of learning and have the opportunities life did not give him. I know in the long ago he once wanted to be a teacher but, forced to earn his keep, he had to take to the world of work and not that of college. Perhaps now I think his words were driven less by anger than by love. He gave me everything he did not have. I pondered all this as I went for my walks in the morning sun of Spain.

Ma and I have spoken on the phone, but she has not mentioned Da or the fight. She tells me the last of the cows calved and delivered a nice bull.

The Spanish and Italians I am with don't know anyone who has delivered a lamb or calf, and ask me how it feels.

'The mother was giving birth – I was there to help, there is nothing to feel, there is no time, instinct kicks in,' I answer.

As when Duncan and I talked, I realize that these facts, this knowledge, now feel as familiar to me as the words of any book.

I am not a typical farmer's son; I left the land so young, but I think now that if that leave-taking had not occurred, I should never have come to see it for what it is: my culture and my birthright.

I look at the animals as so much more than mere beasts. They are creatures of history, holders of the past – our past – and I see in their genes and bodies whole races, not just of cattle but of their owners, the farming families, and in them stories upon stories.

I do not need to be a writer or a farmer. I can be both. I am both.

Future

In the West, the break with farm and butcher is nearly complete. As consumers, we buy most of our meat from supermarkets, packed and sealed. At times it has been dyed with red colorant to make it more appealing to the eye, and at others it has been injected with water by meat packers to add extra weight to increase profits. Many children do not know that beef comes from a cow, and few have seen farms, except on television or in a bedtime story. Most people have never seen a slaughterhouse or a cattle carcass.

So alienated from the living source of our food, it is perhaps inevitable that the next step is to cut out the cow altogether.

In 2016, the Boyalife group opened a facility in China

to produce cloned cow meat. They have already cloned a Tibetan mastiff dog. They say they will not stop at cows and dogs, but will clone cats, racehorses, and even people if the need arises.

They say this technological solution is needed to meet the increasing demand for beef and other meat products in the Chinese market. This development brings with it a host of new ethical questions. Does a cloned animal have a soul? Are we subverting nature by growing meat? Are we prepared to assign the farmer to the growing pile of victims of obsolescence? Are we prepared to end the 10,000-year story of man and cow?

And yet, even as some manufacturers are taking these radical steps towards an artificial future, other farmers are following an alternative path. The organic and grass-fed movement has allowed a small group of growers and farmers to survive corporations, conglomerates and cloners. Organic beef or grass-fed beef may be more expensive to the consumer, but the beast has had a better life, one free of a lifetime of housing, confinement and stress. We raise them to die, but they live a life of peace and nature. Our way of farming here in Ireland, our family's way of farming at Birchview, may be seen as a backward step, but it is a way in which the animal can live with dignity, and one in which the farmer has retaken the old and respectful role of custodian of the land and the environment for the next generation.

There may well come a day when cloned meat is available in the supermarkets of London and the delis of New York, where steaks are grown in labs and test tubes from stem cells, but we have a choice over whether we want this future. Whether we want a cow-less world.

Return

A month passes and my time in the sun ends. I must find the next step in my journey. It is a short plane ride back to Ireland. I have been away four weeks, but it has been long enough to forget, or rather to put aside, hurt and hate. I take the bus home from the airport.

The weather is beautiful and the country is a riot of green; there is life everywhere. With all the animals out in the fields, the yard is silent now, and only Vinny the dog remains to keep watch on the empty sheds. He has grown in my few weeks away and is happy to see me.

That evening, I go for a run through the parish. The northern summer has arrived and the sunlight is long and lasting, and I am glad. I wave to neighbours and friends. I pass Doherty's house in Esker and see the meadow grass growing in the distant fields, I turn left and begin my ascent of the France road and pass Ruske's and Kilnacarrow. The cows, our cows, are in the fields, eating and singing to one another. By Reilly's hills, where they hung the 1798 rebels, I turn and pass the Charters' land. I think of young Willie and his all-too-short life, I carry his memory with me up the road to Gurteen Lake and past its glassy waters. At the village, I take off my T-shirt, for the evening is hot and I run bare-chested now. I bless myself at the graveyard, saluting all the dead: Uncle Mick, Granddad, the Wild Men of Soran, Robin Redbreast. I speed up and move down the hill and wave to Mary as I pass. She is closing up for the evening and smiles to see me half-naked. The statue of Seán Mac Eoin glitters in the fading light and I turn now by our war hero and take the road for home. I cross the bridge of Ballinalee, where it is said a ghost lurks, and peer into the Camlin River, where we built our rafts and spent our summers. I am near the close now and move into our townland of Soran; my breath is heavy and my legs tired, for I have travelled far today. The sweat runs down my back and face and drips into my eyes and I push on and on. By Granny's, I smile and speed up, for I am in the land of the Connells, I am in the galaxy of my home and the universe of my people. I take the right and sprint up our lane. All the way to Birchview farm.

It was eight o'clock when the knock came at my bedroom door.

'I'm going to the bog,' Da says. 'Would you like to come?'

It is the first words we have spoken in over a month and I know this is his olive branch. There is a part of me that seeks to refuse, to break wholly from him. I pause and breathe.

'OK, I'll come,' I say.

The cutting of turf is an ancient thing. The peat is fossilized plant debris, the remnants of ancient forests that once covered the whole island. The turf are cut from the bog and left to dry in the sun. They are then burned in the winter for heat. This work is older than the nation, older than any of us know. The Celts made sacrifices to the bogs, to the gods, and bodies have been found preserved like leather in the peat: men and women of ancient days, their faces contrite, almost prayerful, mummified in the black, gloopy mass.

We work and lift the turf into the trailer. It is black and cool in my hands. He begins slowly to talk to me, telling me his news; he asks of my writing and how was Spain. I tell him there were no cattle but lots of sheep. We talk of football and rugby, of Rory and Uncle Davy. I ask after the new bull and the lambs. We talk now as if we have not spoken in a long time. It has been a long time.

It is as close as he may come to 'I am sorry', as close as we come to 'I love you'. We wipe our brows and curse the heat, but jokingly so, for we are glad of the sun, of the change of season. High above, the birds sing and a murmuration of starlings passes, cresting and falling as a giant black sky shoal. The patterns of nature do not change, but we can.

'The summer is here,' he says.

'Thanks be to God,' I reply.

It is the end of the calving season. We have all our stock, we have each other. It is all we need. It is all we want.

Acknowledgements

This book would not be possible without the help and support of my family. Thank you to my father and mother, who sacrificed so much to give me a good start in life, to my siblings and relations. A special thanks to my grandmother Mary, who, though she lives in these pages, has now departed this life. I hope she would have enjoyed this work. A dear thank you to Father Seán, Liam and the Heslin family, and all those who have walked beside me on the road to becoming a writer. Vanessa from www.writing.ie; Seumas Phelan, Hilary White, Ross Laurence, Jamie and Tim Jones, Elliot James Shaw. A very special thanks to Simon Trewin, my agent; Laura Barber at Granta for her willingness to take a chance and for her wonderful editing and friendship and to Sigrid Rausing for taking a go on my short stories. To Duncan, dear old boy Ramesh, and funny Tar, thank you all for everything and last but not least, Vivian Huynh, the best lady and love I know.

Researching the history of cows was no easy task as there is no standard book that I could find. Many different documentaries, articles and films helped but 'Cowed: The hidden impact of 93 Million cows on America's Health, Economy, Politics, Culture and Environment' by Denis Hayes and Gail Boyer Hayes provided a great launch pad and a helpful insight into in particular modern cattle farming in the USA.

The films of Ken Burns in particular the series 'The West' first broadcast on PBS also helped provide a wonderful insight into the early settler days in the American west and the now famous cattle trails of the nineteenth century. The National Geographic documentary 'Hitler's Jurrasic Monsters' also helped provide more clarity on the rebreeding programmes of Lutz Heck.

And of course the works of Henry David Thoreau which have been a defining moment in my reading life.

Image Credits

p. 19 Wall painting in the Lascaux Cave (UNESCO World Heritage List, 1979), Vezere Valley, France. Source: Shutterstock, image ID 659932630.

p. 22 Auroch by Sigmund von Herberstein. In the public domain via Wikimedia Commons.

p. 40 Egyptian relief with a bull and an ankh, the symbol of life, from Luxor Temple (Thebes), Egypt. Source: Shutterstock, image ID 1084961.

p. 187 Men standing with pile of buffalo skulls, Michigan Carbon Works. Courtesy of the Burton Historical Collection, Detroit Public Library.

p. 209 Heck cattle on Schiermonnikoog island. Source: Shutterstock, image ID 741750736.

p. 223 *Meadow Landscape with Cattle* by Willem Roelofs 1st, *c.*1880, Dutch painting, oil on canvas. Farmer milking cows in the field on the flat landscape. A women stands with a yoke and pails to carry milk. Source: Shutterstock, image ID 379005490.

p. 248 *Melencolia I* by Albrecht Dürer. In the public domain via Wikimedia Commons.

p. 278–9 Author at Birchview. Copyright © Eamonn Doyle/ Neutral Grey.

Keep in touch with
Granta Books:

Visit granta.com to discover more.

GRANTA